MISS DIAGNOSES
MYALGIC ENCEPHALOMYELITIS & CHRONIC FATIGUE SYNDROME

By Byron Hyde MD

Second edition

Dr. John Richardson, In Memoriam

This small compendium of Dr. Byron Hyde's groundbreaking work on Myalgic Encephalomyelitis was collated and edited by Simon Overton, author of "Charcot's Bad Idea". The photograph of the MRI scanner is provided courtesy of Juliet Chenery-Robson. This book was prepared and published on behalf of Byron Hyde MD as a resource for the M.E. community and their physicians. For more information on Dr. Hyde's work see: www.nightingale.ca. This book can be obtained from amazon & lulu.com. This second edition includes Dr. Hyde's chapter from the new Puri & Treasaden textbook "Psychiatry: An Evidence based text".

© Byron Hyde MD 2010

ISBN 978-0-557-32517-7

Contents:

1) *Foreword*
2) *About the author*
3) The Complexities of Diagnosis
4) The Nightingale Definition of M.E.
5) Mental illness in M.E. and Fibromyalgia
6) *Index*

For ten years I had been running a research unit specialising in chronic fatigue and the problems of people who are tired all the time.

- Prof. Simon Wessely

In all M.E. epidemic or endemic patients the patients represent acute onset illnesses. The fatigue criteria listed here [in the CFS definitions] can be found in hundreds of chronic illnesses and clearly defines nothing.

-Dr Byron Hyde MD

A desire to resist oppression is implanted in the nature of man.

-Tacitus

Foreword

Byron Hyde is in the great tradition of those clinicians who compassionately address the needs of their patients especially when their illness(es) are complex and defy prevailing medical fashions and paradigms. He is much sought after by the chronically ill, usually younger patients from childhood to their forties who have benefited greatly from his generosity and willingness to visit them, when necessary, outside his native Canada. He delights in the challenge of clinical examination and accurate diagnosis when faced with complex medical histories and conditions.

For many years he has specialised in treating patients with Myalgic Encephalomyelitis, M.E. a condition first described in 1934 and recognised by the WHO in 1969 as a neurological disorder. The acceptance of Chronic Fatigue Syndrome, CFS, as an alternative name for M.E. in 1988 has given rise to much distortion, dissimulation and deception since chronic fatigue and fatigue syndromes are both classified as mental and behavioural disorders by the WHO. In consequence many patients have had to contend with attempts to diagnose them with a psychiatric illness for which the only treatment is cognitive behavioural therapy and graded exercise coupled, with antidepressants. As a result many patients have become house, or even, bed bound and denied any hope of informed clinical support and treatment.

Byron Hyde believes that too many physicians make an educated but incorrect guess of a patient's diagnosis. He believes that physicians, in the tradition of William Osler, should treat proven pathologies rather than theories that not only are frequently inaccurate, but if the misdiagnosis is a psychiatric diagnosis, then the patient tends to be falsely branded and this incorrect diagnosis tends to be carried over by subsequent physicians.

Byron's response to these examples of poor medicine has been to establish his own Nightingale Foundation and provide a unique service for very chronically ill children, youths and adults poorly investigated and frequently misdiagnosed with a psychiatric diagnosis when in reality they have one or more undiagnosed and significant medically based pathologies. Especially he has drawn attention to-

- The large number of mis-diagnoses of illnesses purporting to be M.E.
- The large number of missed diagnoses, eg thyroid disorders, including cancers.

Byron built on his earlier experience as a geologist and minerals prospector in applying his instrumental skills to medicine. One of his great contributions has come from his careful and perceptive use of technical developments in medicine such as brain SPECT, brain PET, carotid, circulating blood volume and transcranial doppler magnetic resonance imaging, QEEG, and specialised ultrasound studies to provide incontrovertible clarity in diagnosis.

Byron's unique compendium of skills has made him in great demand as an expert witness in medico-legal cases to the relief of many sick people.

For many years Byron has been an active member of the John Richardson Research Group and continued John's tradition of compassionate medicine. In this context he has organised major conferences, particularly Cambridge in 1990, and carried out research studies that provide insights into M.E. The latest is a series of extensive sleep studies identifying extensive deficits in carefully diagnosed M.E. patients.

He has supported political actions by M.E. campaigners in the UK and presented reports to Parliament, especially the Gibson Inquiry. He has a vision of an Institute established to treat M.E. and allied complex chronic disorders.

It is a privilege to know him as a physician, and colleague but above all as a friend.

Malcolm Hooper (Prof. Emeritus of Medicinal Chemistry, Sunderland) 16 April 2009

About the Author

Dr. Byron Marshall Hyde studied pre-medicine in the Faculty of Medicine, University of Toronto followed by a degree in Chemistry and Nutrition in 1961. His first medical employment was as an immunological research chemist at the Roscoe B. Jackson Laboratory, Bar Harbor, Maine - a leading world laboratory in immunological & transplantation research. He then became Chief Technician in charge of the Electron Microscope Laboratory at the Hospital for Sick Children in Toronto.

Dr. Hyde returned to the University of Ottawa and graduated from the Faculty of Medicine in 1966. After an internship at Montreal's Hotel Dieu and residency at the St. Justine Paediatric Hospital and the Ottawa Civic Hospital, he opened a family practice in Ottawa that continued until 1984 when he started the full time study of post infectious Myalgic Encephalomyelitis. For five years he had travelled extensively around the world investigating the epidemics of M.E. in the USA, the UK, Australia, New Zealand and Iceland and spent the next several years being instructed by previous researchers of these epidemics. Only then did he start to investigate patients who had M.E.

In order to widen resources to investigate these patients, in 1988, he founded the Nightingale Research Foundation, obtaining charitable organization status in the same year. Nightingale is dedicated to explore, understand and treat the patients disabled with Myalgic Encephalomyelitis, Chronic Fatigue Syndrome (M.E. and CFS), fibromyalgia-type illnesses and post-immunization injuries. In its early years, Nightingale became a critical vehicle providing technical assistance to other medical practitioners and researchers worldwide and outreach and informative publications to help and encourage thousands of North Americans who were patients or had family members disabled by M.E. or CFS.

In the 1980s, little physician or patient-based research existed into these poorly understood illnesses. At that time few North American physicians were unaware of the excellent UK diagnostic criteria and clinical definitions of M.E. In the UK, the excellent work of many of the early M.E. physicians had come to a virtual standstill and there was an increasing attack on both the old and new M.E. physicians by many psychiatrists and a general lack of knowledge by the physicians. In Canada, medical research had no appreciable funds to unravel the enigma of these illnesses and government cutbacks only added to the difficulty of finding funds for investigation of M.E. and CFS patients. Nevertheless, Dr Hyde, with the help of a dedicated band of volunteers and physicians, set out to change this situation that was causing even greater injury to the already disabled M.E. patients. Thanks to the assistance of many medical practitioners and scientists, investigational research slowly advanced until by 1998, Dr Hyde was able to diagnose the cause of the M.E. and CFS type illnesses in 90% of the patients who came to his office for investigation.

As he expanded his knowledge of this group of diseases, he collected critical comparative data for Nightingale's patient research database. Consequently, in 1990, in collaboration with Dr. Richardson and the Newcastle Research Group in the UK, the funds were raised to organize and convene the First World Symposium on M.E. and CFS at Cambridge University in the UK. Dr. Hyde went on to collaborate with over 100 experts to edit and publish in 1992 the 725-page 1992 encyclopaedic textbook, "The Clinical and Scientific Basis of Myalgic Encephalomyelitis and Chronic Fatigue Syndrome", the first comprehensive and authoritative medical reference book on M.E. and CFS still widely cited by researchers internationally.

Initially, many physicians were not aware of the fact that many children fall ill with M.E. and CFS-like diseases. Dr Hyde still spends considerable time assisting children and students with M.E. and CFS. In part, due to this interest, Dr. Hyde noted that acute onset M.E and CFS patients had similar SPECT brain scan topography to children he had examined with acquired autistic type injury. In 1989, he made this statement to a San Francisco medical symposium that included Dr. Michael Goldberg, a paediatrician from Tarzana, California who had noticed a similar spectrum of brain changes in his autistic children patients. Dr. Goldberg started the Neuro-Immune

Dysfunction Syndromes (NIDS) association with Dr. Hyde as one of the founding board members. This joint pioneering work with children afflicted with this youth-robbing illness spectrum established a world-wide friendship network of doctors interested in collaborating in this field of autistic and acquired brain dysfunction research.

Dr Hyde re-grouped Nightingale and focused its efforts in 1995 on collaborative research with like-minded medical practitioners and researchers internationally and on its publishing activities. Today, Nightingale's priorities are individual patient-based research with total body / brain investigation of M.E. and CFS patients as well as the development of a sophisticated database to consolidate these findings for analysis and publication. The uniqueness of his work is in its emphasis on total body mapping of all systems and organs so that he can understand the nature and complexity of the M.E. and CFS patients' illnesses. He is one of the few physicians worldwide whose practice has consisted solely of the investigation of M.E. and CFS patients since 1984.

Dr. Hyde is also active in individual case M.E. and CFS research, providing advice on legal and medical research issues in order to provide moral and scientific support for the small but growing number of leading edge M.E. and CFS diagnostic initiatives. Today, he probably manages more detailed data base information in this area of inquiry than any other researcher in North America as a result of his investigations into acute onset and gradual onset central nervous system dysfunction associated with fatigue and pain syndrome. Depending on adequate funding, it is planned to update this database composed of input from over 3,000 patients, and publish various epidemiological studies of this population as well as other publications and journals of interest to the CFS and M.E. community. Recently, this detailed patient investigational work has demonstrated new factors: the high rate of thyroid malignancy that exists in patients who initially fell ill with chronic fatigue syndrome. Thyroid cancer exists in only 1 per 100,000 of the general public, yet 6000 per 100,000 M.E. and CFS patients. These results have been accepted for publication in a leading nuclear medicine journal and will be made available on the

Nightingale website.

Nightingale has a new database program that has been generously funded by its patients and their families. It is at the heart of a second level of clinical research based on all of Dr Hyde's investigational patients. For many of these patients, this investigative data covers over 20 years and is proving to be one of the best M.E. and CFS longitudinal databases in existence. In addition to the thyroid pathologies noted above, these data have already begun to yield discoveries not found elsewhere. Dr. Hyde believes that the vast majority of the gradual onset group of CFS patients was misdiagnosed and many have subsequently been found to be suffering from major organ or system pathologies that had not been identified by primary care physicians or subsequent specialists.

Dr. Hyde continues to practice medicine in Ottawa. He has published a number of poetry and prose works. Dr Hyde is married to Lone Due Petersen and has eight grandchildren and the family dog Atimus and the cat called Maus.

The material in this chapter is used by Simon Overton with the permission of Dr. Byron Hyde and Dr. Leonard Jason to reprint under written permission to Dr. Hyde from John Wiley & Sons Inc., 111 River Street, Hoboken, NJ 07030, USA from Vanessa Miranda, Digital Rights Coordinator on 29 July 2005. Copies of the Jason-Fennell-Taylor textbook, Handbook of Chronic Fatigue Syndrome, ISBN 047141512X can be obtained from the Publisher, www.wiley.com

The Complexities of Diagnosis

INTRODUCTION

SINCE 1985, I have restricted my practice to the investigation of patients with myalgic encephalomyelitis/chronic fatigue syndrome (M.E. /CFS) and the underlying causes of their illnesses. Whenever possible, I spend an entire day and part of the next examining the patient, and this has allowed me to observe anomalies that might elude other physicians. I then embark on a systematic mapping of the patient's structures, systems, and organs. Only then do I reach a conclusion about the disease process that I am trying to understand and uncover. Few physicians have such a luxury of time. Because this protocol is costly for the medical system under which physicians generally practice, I do not suggest that the readers of this chapter must follow it. It might lead to financial difficulties if the physician is a fee-for-service physician or practices within a medical group that dictates the amount of time to be spent on consultations.

My methodology has produced certain views and findings that may be inconsistent with many in this field. The opinions expressed here are, in large part, a result of my own clinical experience and are not derived from empirical findings from controlled research studies. At the Nightingale Research Foundation, Bonnie Cameron, Lydia Neilson, and I did a survey of 2,000 patients in 1990 - 1994 that has informed many of my opinions, but the study was not published.

INVESTIGATION OF M.E. /CFS

In our text, *The Clinical and Scientific Basis of Myalgic Encephalomyelitis/Chronic Fatigue Syndrome* (Hyde, Goldstein, & Levine, 1992), I distinguished between M.E. and CFS. More recently, and specifically at the biennial CFS Symposium held in Seattle in 2001, some individuals, dissatisfied with the name chronic fatigue syndrome, suggested changing it to myalgic encephalomyelitis or some variation of that name.

This would be unwise. Although M.E. and CFS share many characteristics, the titles often represent two distinct groups of illnesses.

MYALGIC ENCEPHALOMYELITIS (M.E.)

The term *myalgic encephalomyelitis* was based on clinical descriptions of an illness that has occurred both sporadically among the general population and in clusters, or epidemics, usually in hospitals or schools. Over 60 such epidemics have been described in the medical literature (Acheson, 1992; Henderson & Shelokov, 1992; Hyde, 1992) since Sandy Gilliam, Assistant Surgeon General of the United States and later Dean of Medicine at Johns Hopkins, first described the 1934 epidemic in the Los Angeles County General Hospital (1938). B. Sigurdsson et al. (1950) in Iceland, D. A. Henderson, and A. Shelokov, in the United States (1959a, 1959b); A. Wallis, in 1955, and A. M. Ramsay, in 1988, and John Richardson, in England (1992); and P. Behan, in Scotland (Behan & Behan, 1988; Behan, Behan, & Bell, 1985), have all added to this growing literature. This group of illnesses has been given many names, but these have distilled down to myalgic encephalomyelitis (M.E.), a term used primarily in the United Kingdom, Canada, and Australia.

These various clinical descriptions include these characteristics:

- A sporadic and epidemic postinfectious illness most frequently occurring in the late summer or early autumn, with an incubation period from 4 to 7 days. The epidemic illness is most commonly acquired in hospitals, schools, or domiciliary institutions at a time when an increase of similar sporadic illness occurs among the general population. Although the illness is seen in diverse occupations, health care workers, teachers, and students are the most commonly affected.
- The epidemic illnesses have been associated with infrequent deaths involving CNS (central nervous system) changes. Many of

these changes have been subcortical brain changes. Deaths in sporadic cases have been rare but have been associated with acute cardiac arrest, with no signs of coronary disease, and frequently suicide. Deaths other than suicide are uncommon.
- Onset of the primary M.E. illness usually follows abruptly during the recovery phase of an often banal infection (if an infection is noted at all) or within 4 to 20 days of an immunization. Frequently one observes the onset of an M.E. -like illness after multiple infectious episodes. The primary infectious illness and the M.E. illness do not resemble each other. Most infectious illnesses are described as upper respiratory tract, flu-like, gastrointestinal and, less commonly, hepatic illness or pneumonia. Traumatic incidents associated with minor infectious illness or travel to foreign countries. These associations often follow within 30 days of a series of immunizations.
- M.E. illness in adults is associated with measurable changes in the CNS and autonomic function and at times injury to the cardiovascular, endocrine, and other organs and systems. It is described as (1) a systemic illness often of subnormal temperatures; (2) marked muscle fatigability; (3) an acute onset of CNS changes of memory impairment, mood changes, sleep disorders, irritability, and reactive depression; (4) involvement of the autonomic nervous system resulting in tachycardia, coldness of the extremities, urinary frequency, bowel changes, pallor, and sweats; (5) diffuse and variable involvement of the CNS leading to severe headaches, visual problems, ataxia, weakness, cramps, and sensory changes; (6) muscular and neck pain, acute fleeting spasmodic pain and tenderness, and myalgia.
- In children in the acute phase, there is depression with weeping, significant loss of energy, retardation and impairment of thought and memory process, disorders of sleep, behavioral disorders, acute onset of school problems, often of a serious nature, with a reluctance to attend school, and with a significant weight loss. Children are usually diagnosed as hysterical or school phobic.

- The initial period of illness lasts from weeks to up to two years and tends to be more severe. During this period, the patient either recovers, remains, or relapses in a chronic phase of variable severity. The chronic phase is often sufficient to prevent return to school or work for either long periods or permanently.
- Dr. Michael Goldberg, of Tarzana, California, believes that this illness often results in children being rejected, abused, and abandoned to the street or to juvenile criminal activity. Dr. John Richardson (1992), of Newcastle, and others have documented significant associated cardiac and cardiovascular injury as well as other organ injuries associated with the usual CNS and autonomic changes in this group of patients. Dr. Seymour Grufferman (1992), of Pittsburgh, has described an increased incidence of malignancies, often lymphomatous, associated with individuals in clusters of M.E. /CFS. A similar finding was initially described in some of the patients in the Lake Tahoe epidemic (Daugherty et al., 1991; Peterson et al., 1992).

All M.E. descriptions were concerned with chronic or recurrent acute onset illnesses. The M.E. descriptions deal with primarily CNS and autonomic changes and, at times, with easy fatigability and with poor or delayed recovery of CNS or muscular abilities. Although M.E. clinical descriptions noted the infectious onset and infrequently the postimmunization history of M.E. illness, neither pharyngitis nor involvement of lymph nodes was ever mentioned in any of the clinical descriptions of the actual chronic illness manifestations.

Host factors are important in M.E. . At the time of the initial illness, the patient often appears to be either temporarily or chronically immune-compromised by one or more of the following:

- Exhaustion from overwork or night shifts
- Repetitive infectious disease
- Recent immunization
- Significant illness or trauma
- Toxic chemical exposure

As in all diseases, there is a significant variation in the degree and range of injury. Those who are least injured often simply return to school or work and operate at a lower productivity and escape diagnosis. Those who are most injured or die are easily recognized at disease onset or shortly after as CNS, cardiovascular, or organ injury. Because of their overwhelming illness and the specificity of the end-organ injury, they are never diagnosed as M.E. except in epidemic or cluster situations. Overwhelming fatigue is often a feature of the chronic illness phase (Fukuda et al., 1994). After a few months, however, this profound fatigue often changes and some patients begin to feel normal until they are challenged by any physical, intellectual, emotional, or sensory stress. In this new phase, the patient has rapid fatigability and poor recovery after any stressor. These patients begin to feel they inhabit a body and mind significantly different from usual, and sometimes they panic. The adult patient with moderate to major illness rarely recovers totally, but usually does improve (Joyce, Hotopf, & Wessely, 1997). It is an unacceptable improvement. Those adults who are still significantly ill at two years can still improve but only a few ever return to any degree of normal function. Unlike adults, the majority of children and adolescents, even those seriously injured, who have proper care and are in a positive economic environment, tend to recover substantially or at least improve significantly over time (Marshall, 1999).

CHRONIC FATIGUE SYNDROME (CFS)

The physician and patient alike should remember that CFS is *not* a disease. It is a chronic fatigue state as described in four definitions starting with that published by Dr. Gary Holmes of the CDC and others in 1988 (Holmes, Kaplan, Gantz, et al., 1988; Holmes, Kaplan, Schonberger, et .al., 1988). The definition created by Lloyd, Hickie, Boughton, Spencer, and Wakefield (1990) is also widely used in Australia. There are two subsequent definitions. The Oxford definition of 1991 (Sharpe et al., 1991) and the 1994 NIH/CDC definitions (Fukuda et al., 1994) are basically, with a few modifications, copies of the first definition. Where the one essential characteristic of M.E. is acquired CNS dysfunction, that of CFS is primarily chronic fatigue. By assumption, this CFS fatigue can be acquired abruptly or gradually. Secondary symptoms and signs were then added to this primary fatigue anomaly. None of these secondary symptoms is individually essential for the definition and few are scientifically testable. Despite the list of signs and symptoms and test exclusions in these definitions, patients who conform to any of these four CFS definitions may still have an undiagnosed major illness, certain of which are potentially treatable. Although the authors of these definitions have repeatedly stated that they are defining a syndrome and not a specific disease, patient, physician, and insurer alike have tended to treat this syndrome as a specific disease or illness, with at times a potentially specific treatment and a specific outcome. This has resulted in much confusion, and many physicians are now diagnosing CFS as though it were a specific illness. They either refer the patient to pharmaceutical, psychiatric, psychological, or social treatment or simply say, "You have CFS and nothing can be done about it."

The CFS definitions have another curiosity. If in any CFS patient, any major organ or system injury or disease is discovered, the patient is removed from the definition. The CFS definitions were written in such a manner that CFS becomes like a desert mirage: The closer you approach, the faster it disappears and the more problematic it becomes.

SIGNIFICANT DIFFERENCES BETWEEN M.E. AND CFS

Though the symptoms of CFS resemble those of M.E. , the differences are so significant that they would exclude M.E. patients from the 1988 and 1994 CDC diagnoses of CFS. The following features of M.E. separate it from CFS:

- The epidemic characteristics
- The known incubation period
- The acute onset
- The associated organ pathology, particularly cardiac.
- Infrequent deaths with pathological CNS changes.
- Neurological signs in the acute and sometimes chronic phases.
- The specific involvement of the autonomic nervous system.
- The frequent subnormal patient temperature.
- The fact that chronic fatigue is not an essential characteristic of the chronic phase of M.E.

However, there are four essential differences between M.E. and CFS that are perhaps more important than any of the preceding differences:

1. No one in composing the two CDC definitions told anyone not to investigate the CFS patients during the first 6 months of illness; they simply stated that the CFS is characterized by an illness of 6 or more months of chronic fatigue. Undoubtedly, it was unintentional. Yet obviously CFS following infectious disease begins in day one of the first 6 months or even in the days before this initial period. Researchers into CFS have simply avoided that essential area. The inception of an illness is always the most fertile area of research into cause and pathology.
2. Organ disease in CFS has been avoided. By definition, it does not occur. If significant primary or secondary organ disease occurs, then this would be a cause of the fatigue and the illness would not be CFS (Fukuda et al., 1994).
3. The inventors of the second CDC CFS definition laid out certain guideline examinations (Fukuda et al., 1994). They never stated that no other testing should be done, but for all purposes, these very preliminary tests have been used for inclusion guidelines in CFS research papers. Research physicians have apparently forgotten that we do not know what CFS is from a

pathophysiological basis. For this reason, not only have most physicians avoided exhaustive testing but many have decried exhaustive testing as foolish.
4. This is the most important essential difference. Nowhere in any of the four definitions of CFS is there a discussion of acute versus gradual onset illness. This has allowed physicians to include any patient who fits the 1988 or 1994 or U.K. definitional characteristics into the CFS illness spectrum. Because none of these definitions mentions gradual onset CFS disease, gradual onset patients, as a group, not only fit the four definitions but also totally obstruct CFS as a disease category. The reason for this statement is simple. Gradual onset CFS frequently represents non-diagnosed major disease or pathophysiological anomaly. Many patients with a diagnosis of CFS today have non-diagnosed major diseases. These patients warp any statistical or scientific examination of the CFS patient. Most of the patients I have seen from Canada, the United States, or from the United Kingdom with gradual onset CFS illness have non-diagnosed major medical illness or anomaly. This fourth essential difference defines the cornerstone of investigation of much CFS.

PREMISES CONCERNING THE PATIENT AND THE DISEASE ENTITY

The patient with the diagnosis of M.E. /CFS is chronically and potentially seriously ill with (1) a poorly understood illness of a pathophysiological nature or (2) a missed classical disease entity. The typical patient has seen many excellent physicians, who have failed to discover the cause of the patient's illness other than to variously call it M.E. or CFS, psychiatric illness, somatization, or more charitably, "I simply do not know." These physicians have repeatedly performed many tests but have generally failed to find any significant or substantial indication of cause or nature of the patient's disease.

At least some of the patients with an initial diagnosis of *gradual onset* M.E. or CFS have another and potentially treatable classical disease or anomaly. These M.E. /CFS patients require a total investigation and essentially a total body mapping to understand the pathophysiology of their illness and to discover what other physicians may have missed. In many instances, patients appear to know more about M.E. /CFS than their physician and in fact have directed their own investigation under the directional guidance of a kind and supportive clinician.

These patient-directed investigations usually jump from one trendy test of little value to another consuming vast amounts of funds and time. Rarely, however, do the physician and patient end up with any substantial scientifically supportable disease entity or diagnosis other than that with which they started – M.E. /CFS. One can assume that many of the patient's physicians have spent the proverbial 8 minutes that an average North American or British physician spends with the average patient. Likewise, most internists will have spent 40 minutes doing a classical history and physical that can generally detect obvious acute disease or advanced disease of a progressive nature, but is usually irrelevant in understanding a chronic pathophysiological illness.

I assume that none of the patient's illnesses is due to a psychiatric cause until I have completed my investigation. In the end, although these patients may have significant anxiety and problems caused by loss of income, social status, and meaning, less than 5% have any significant psychiatric illness. Initially in 1985 to 1990, I was able to unravel the causative disease or illness in the M.E. /CFS group in no more than 10% to 20% of the patients I examined. By 2000, I was able to discover the major elements of the underlying disease pathophysiology in 70% to 80% of the patients I examined. Each year, my success ratio has improved. Because of this, I believe that the 20% to 30% failure rate in defining the pathophysiology of this group is due to my own deficiencies as a physician and/or the deficiencies of the available technologies. One should not blame patients for their illness or jump too casually to a psychiatric or sociological diagnosis.

For me, a patient with an initial diagnosis of M.E. /CFS can be a gold mine of disease, missed injuries, physical and physiological anomalies, and genetic curiosities.

PHILOSOPHY AND ECONOMIC ETHICS OF INVESTIGATION

I base my philosophy of examination and testing of M.E. /CFS patients on the following considerations:

- The majority of M.E. /CFS patients who seek medical assistance in my practice tend to be middle-income individuals or professionals. Many have been unable to work for years. The patient's loss of income for one year usually represents more than $30,000, and I have seen patients who have lost an income in excess of $500,000 per year. These individuals tend to range in age between 20 and 40 and are in the prime of their work life when they first fall ill. If they cannot return to work, the gross income loss to themselves, an insurer, or the state – or simply the loss of their productive life – can often reach $1 million to $15 million.
- The technological component of a reasonably complete investigation and body mapping rarely should cost more than $10,000. The term *body mapping* is an idiom I adapted from another profession. Prior to becoming a physician, I worked as a geophysicist, and to evaluate an anomaly, it was necessary to first map the terrain in detail with surveying and geophysical tools to measure the size, depth, and nature of the anomaly. Diamond drilling and core analysis often followed the initial measurements. I helped discover several mines that people had often walked over without even realizing what they had missed. The investigation of an M.E. /CFS patient is similar to the research leading to the discovery of a gold or nickel mine or an oil deposit. Before you can know and understand what anomaly you are dealing with, it is

sometimes necessary to do a total body mapping. This may cost approximately $10,000. At this point, many physicians and insurers will throw up their hands in exasperation. They cannot justify spending such a considerable sum to investigate a patient who, in their judgment, has an obviously psychiatric/somatic illness or is simply too lazy to return to work and has been properly investigated by several reputable physicians.

- The majority of M.E. /CFS patients cost themselves, or the medical system in which they operate, far more than $10,000 over the course of their illness in a totally nonstructured series of haphazard investigations. Even compared with a year's income, $10,000 represents a fraction of the patient's or employer's loss. When compared with a lifetime loss of $1 million to $15 million, such an investigation cost is paltry.
- Although individual patients may know a great deal more than I do about some particular aspect of some particular disease or pathology, they are not physicians. Nor is the patient usually trained as a physician, and rarely does the patient understand the rigors of scientific medicine. Some physicians, in attempting to earn a reasonable living in these economically challenging times, have also relinquished their investigational skills. It is not for nothing that medical training is so prolonged. Medicine is a difficult profession; medical investigation is a difficult pursuit; the investigational physician can never know enough or have enough tools or instruments to measure everything that needs to be measured. Faced with the challenge of M.E. /CFS illness, no physician can presume to understand the pathophysiology of this group of illnesses; particularly after only an hour with the patient and a few standard tests.
- The chronic M.E. /CFS patient deserves, at least once, a complete investigation that includes mapping of (1) body structures, (2) organs, and (3) systems. Where little or nothing is initially discovered, the same physician should repeat this investigation after a few years. Over time, even chronic disease tends to be progressive and more visible to investigation. Also, physician skills

and professional knowledge continue to improve. Patients routinely arrive in my office telling me they have had a complete workup, but few of these patients have had what I consider to be even basic investigation.
- The investigation of a known illness such as heart valve disease or a brain tumor is relatively simple and can be completed with an economy of tests and examinations. A chronic disease process that is poorly defined or is of unknown origin requires a different approach.

THE PHILOSOPHY OF TREATING M.E./CFS DISEASE

Though M.E. /CFS usually represents significant disease processes, the underlying pathophysiologies or physical anomalies causing these processes are so varied that it is unreasonable and perhaps even dangerous to suggest or embark on any uniform treatment.

Although CFS has been defined as a syndrome, patient, physician, and even government agencies have increasingly tended to speak about CFS as a specific disease entity with a potentially specific treatment or treatments. Whether this suggested treatment protocol employs pharmaceuticals, cognitive or physical retraining, or alternative medications and treatments, these treatment modalities and philosophies are not medically justifiable and are often potentially dangerous to the patient.

In the past two centuries, the development of Western medicine was based on autopsy, physiology, pathology, and reproducible tests. The goals were to define and, where possible, treat the causes and/or the pathophysiology of the disease process. This philosophy of modem Western medicine has been the basis for almost all of the great medical cures and treatments for specific diseases during the nineteenth and twentieth centuries. To date, however, this approach has largely been missing in the investigation and understanding of M.E./CFS disease.

There has been an immoral intervention by the insurance industry into the philosophy of physicians and health workers treating this group of disease entities. This corporate insurance company intervention has used the mechanism of sponsoring medical symposiums to produce a uniform, insurance-friendly policy. Insurance companies have reputedly placed large numbers of rheumatologists and specific subspecialists in a given area under a significant annual retainer, injuring not only patient access but also negatively influencing other physicians who may not be aware of this economic relationship.

INFLUENTIAL FACTORS IN TREATMENT

The definitions of myalgic encephalomyelitis (M.E.), chronic fatigue syndrome (CFS), and fibromyalgia have colored all investigations of this illness group. The definitions of myalgic encephalomyelitis and chronic fatigue syndrome describe what may originally have been the same disease, but the differing definitions have caused confusion.

FIBROMYALGIA AND VASCULAR PAIN

Both M.E. and CFS patients may have associated pain that includes fibromyalgia (Taylor, Friedberg, & Jason, 2001). Some have no associated pain dysfunction. The pain syndromes, and there are many, vary in intensity and tend to be worse in the first years of illness and after the patient has encountered physical, intellectual, sensory, or emotional stressors. Although some researchers have found specific chemical changes in the spinal fluid of these patients, and others have demonstrated subcortical SPECT(single photon emission computed tomography) anomalies (Goldstein, 1992), it is likely that in the future measurable findings may be found in the posterior columns and posterior root ganglia. If physiological spinal cord changes occur, they have not been subjected to scientific scrutiny because specific noninvasive testing modalities are not yet available.

Instead of following neurological pathways, some of these pain mechanisms are probably vascular. If this is true, this may suggest injury to the autonomic system. Any physician subjecting this category of patient to a thallium chemical cardiac stress test will know that many of these patients experience severe incapacitating pain that sometimes lasts for days, even several weeks. Although I do not know if there is a CNS or spinal basis to these pain phenomena, the paradoxical thallium test would suggest a vascular basis to the pain dysfunction in this group of patients.

Raynaud's phenomenon is a common secondary occurrence in both M.E. /CFS and fibromyalgia. When significant fibromyalgia occurs in conjunction with M.E. or CFS, the chronic disability tends to be additive.

ACUTE AND GRADUAL ONSET ILLNESS

I tend to arbitrarily place acute onset patients in the M.E. category and the gradual onset in the CFS category. This arbitrary categorization is not entirely satisfactory. Some patients have no idea if their illness started abruptly or gradually. Even so, overlap in these two groups makes it an imperfect analysis. Remember, a patient with M.E. is a patient whose primary disease is CNS change, and this is measurable. The primary disease of a patient with CFS is fatigue, and fatigue is neither definable nor measurable.

The gradual onset CFS group is of particular concern to me. It is in this group that occult disease, whether malignant, space occupying, organ pathology, or vascular injury of the CNS or cardiac system, is most frequently observed. A typical M.E. -like history can often be due to a malignancy (Richardson, 1992) or other pathology that should be located as soon as possible. Whether a patient fell ill abruptly or gradually, or has been ill for many years, is no excuse not to search for a potentially treatable malignancy or a cardiac, vascular, or other organ illness. Patients with M.E. /CFS are not immune from developing other illnesses that may be potentially terminal.

THE RATIONALE FOR INVESTIGATION

Scientific Medicine   The tradition of Western scientific medicine is to isolate the cause of the illness, measure it, and specifically treat that cause if possible. Without being able to understand and measure the nature and degree of the underlying injury or disease, it is impossible to measure the effectiveness of any treatment. Some causes of M.E. /CFS-like illness are eminently treatable, and effective treatment may allow the patient to go back to work or school.

Understanding   Patients want to know and have the right to know what has happened to them.

Insurance Indemnity   Some, if not most, insurance companies do not accept the diagnosis of M.E. or CFS as a basis for disability even if the patient is permanently bedridden or confined to a wheelchair. The physician must be able to demonstrate the underlying injury to a court, if need be, to assist the truly disabled patient in claiming a disability pension. Although this does not treat the disease, at times it can materially restore the disabled individual to acceptable financial stability, without which life often becomes intolerable.

Financial and Social Loss   The majority of patients with M.E. /CFS-type illness tend to be professionals or individuals with an above-average education and a successful career, who may forfeit significant income because of work loss in their lifetime (Anderson & Ferrans, 1997). Each of these patients requires a complete clinical, laboratory, and scientific investigation at least once. Although I have seen some patients who were charged $20,000 to $100,000 for $3,000 worth of tests, a complete technical investigation should cost less than $10,000. Many physicians and corporate organizations think that the state or their company cannot afford to investigate a patient in such depth, even though the state or insurance company has no difficulty in paying that patient $10,000 to $20,000 a year or more in social benefits if the disability is accepted. A thorough evaluation of the patient could help eliminate this problem of rational accountability.

THE IN-DEPTH EVALUATION

PATIENT HISTORY

In addition to the regular history, prior to the first visit, I have the patient provide a full extended family genealogical health history going back three or even four generations and including siblings of each generation, all of their known illnesses, and cause of death. Patients who know their birth parents usually can obtain this history. Frequently, mapping this genetic history suggests or even reveals the source of the CFS patient's real illness. Patients with M.E. sometimes have a curious history. I often find an excess of recurring and major neurological illnesses in previous generations. Even though paralytic poliomyelitis was relatively rare, it is common to find one or more polio victims in the family tree. I have often wondered if these patients do not suffer from a specific immunological dysfunction to neuropathic viruses.

GEOGRAPHIC HISTORY: A POTENTIAL INVESTIGATION BLIND SPOT

Patients, as they should be, are very concerned about toxic chemical exposure as a cause of their illness. Physicians and governments pay lip service to this concern, but perhaps because of lack of technology, I have found little supportive evidence to substantiate toxic chemical exposure as a cause for chronic M.E./CFS (Crowley, Nelson, & Stovin, 1957; Shelokov, Habel, Verder, & Welch, 1957). However, patients who have a history of being raised on an active twentieth-century farm or a village with no central water supply are potentially victims of well water toxic chemical exposure.

Well water is normally only routinely examined for bacteria, and in my experience, farm and village well water is almost never examined for pesticides or herbicides. In villages and towns that rely on local or central well water, whose source is near a major farm area, chemical factory, or dry cleaner, the users of this water may have been subjected to toxic chemical exposure in their water for decades. This type of exposure can lead to a gradual immune breakdown. Hair or serum analysis may not demonstrate these old exposures. Many of these toxins are lipophilic and a fat biopsy (liposuction) should be considered in this group. Since pesticides, herbicides, and all organophosphates accumulate in fat and the brain is essentially fat, the brain should be considered to be a natural reservoir of these chemicals. The brain, of course, is also the major immune-regulating organ of the body. Since we cannot routinely do brain biopsies, analysis of samples of liposuction fat may help to identify toxic levels of these chemicals.

PSYCHIATRIC HISTORY

Patients frequently hide their family psychiatric history. Often I am assured that there is no psychiatric history, and then after an exhaustive examination, I find no physical causes for the patient's illness. Patients with M.E./CFS with no observable pathology are very infrequent. I then go back and ask specific questions of each family member. Infrequently, I find a severe psychiatric history. Having said this, I doubt if more than 2% to 5% of M.E./CFS patients have a primary psychiatric history. Why should there be fewer psychiatric patients in this group? The patients that I see, particularly in the acute onset group, are primarily professional middle-class individuals. They have worked hard for years to further their careers, and most persons with major initial psychiatric illness would have simply failed to achieve this success.

PHYSICAL EXAMINATION

The severity of M.E./CFS illness is not usually accompanied by significant observable physical changes in the regular physical examination. This causes some physicians to assume that there is no major disease present in patients with M.E. /CFS. Yet most male patients I have seen have never had a rectal or women a vaginal exam, and almost none have ever had anyone look into their nasal passages.

The physical examination does not start on the examining table. My physical examination often starts with the moment the patient gets out of a car to come into my office. Severely ill patients seldom come since they cannot get out of bed or handle the nine steps to my office. These patients I see only in their homes. The mild cases, who keep working or going to school, are rarely if ever seen by a physician. I have an advantage in often being able to see the patient from my window. Typical moderate to moderately severe M.E. patients often cannot get out of a car normally. Patients who have driven any distance often physically lift their legs out, first one leg then the other, hold onto the car door frame, and struggle out. Their upper leg flexors are unusually weak. Sometimes they have short-term foot drop and cannot raise their foot. I have a simple gate catch that a child can open instantly. I have seen patients work for several minutes without being able to open the latch. A sign on my outside door says "ring and enter," and often the patient simply rings and stands there. Once inside, the patient often goes down the steps one at a time, one foot leading while holding firmly on to the banister.

During the daylong examination I often accompany patients for tests in the hospital simply to observe them. They frequently do not walk normally; they get lost in their purses or wallets attempting to find their identification. Walking with these patients is often like walking with a tortoise. They can be slow, clumsy, sometimes walking with a wide leg stance. Some have a movement disorder that does not conform to the classical Parkinson or upper motor neuron disorders. These patients have obvious CNS injury but simply do not fit into neat categories.

Initially, patients are often excited about seeing me, their adrenaline pumping, and a physician who saw some of these patients for only up to an hour would reasonably conclude that they were high-energy patients with nothing wrong. This is misleading. During the course of a day's examination, the patient may change from a brighter than normal person to one who resembles a blank-faced zombie, a patient who can talk and walk only with difficulty or not at all. Sometimes their voices become scanning, and they begin verbally to stumble. Normally, I take the patient to lunch. This helps me diagnose the infrequent bulimics. Sometimes patients are fine all day, but when I see them on the second day, they have often, in physical and intellectual terms, gone to pieces. A one-hour physical examination will rarely pick up M.E./CFS pathology.

Oral Temperature   Prior to the office visit, I have patients take a temperature reading at specific times, 4 times a day for three days, and also ask them to have a healthy friend of the same age and sex provide a similar temperature series for comparison. This is not a good test due to the variation of procedures and menstrual cycles, but the patient with acute onset M.E./CFS frequently has a substantially subnormal temperature. In 15 years of examining chronic M.E. or CFS patients in Canada, the United Kingdom, and Australia, and in CFS clinics in the United States, I have found an elevated temperature on only two or three occasions. The significance of elevated temperature in the CFS definition eludes me. Patients have subnormal or normal temperatures.

Cervical and Axillary Glands   The initial CDC case definition for CFS suggested as a physical criterion, "Palpable or tender cervical and axillary lymph nodes" (Holmes et al., 1988). Few of the signatories of that definitional paper were actually clinicians who had ever seen any M.E./CFS patients on a regular basis. The Oxford Group corrected that and simply stated, "There are no clinical signs characteristic of the condition" (Sharpe et al., 1991, p. 119).

M.E./CFS patients frequently have surface hypersensibility or pain syndromes, but since 1985, I have rarely found significant cervical or axillary glands in an M.E./CFS patient. Sometimes they do have painful elliptical swellings. When they occur, they can be quite large but are fleeting. One is located above and to the right of the left mammary gland. Often if you ask, the patient will go to that point but say it "isn't there today." Over the years, several M.E. /CFS patients have told me that they have had their left breast biopsied at this exact site for possible malignancy and nothing was found.

Another location is a row of these elliptical swellings in the left axilla at the chest wall muscle edge. These may come out during the first few years of illness at any time and later when the patient is tired. They are subcutaneous and tender and in severe cases cause a bruise or discoloration over the spot. You can roll them under the ball of the finger. They are exquisitely painful. They are never constant.

In the past, I have had two patients biopsied and found that there were no abnormal lymph nodes but a bundle of histiocytes. Dr. J. Gordon Parish demonstrated to me that often the subcutaneous anterior upper legs are also "lumpy" in these patients. If you find enlarged lymph glands in the cervical or axillary areas, look for other causes than M.E. /CFS.

OTHER PHYSICAL FINDINGS

Of the following 20 abnormal findings in these patients, none is strong enough to excite most internists or neurologists. The findings are not usual in a healthy patient, however, and many are not specific to M.E. /CFS. There tend to be more findings in the early illness, but some persist and appear to increase during the course of the day:

1. Ghastly pallor of face with frequent lupus-like submaxillary mask
2. Parkinsonian rigidity of facial expression and altered walk
3. Scanning, disjointed speech, or reversals
4. Nasal passage obstruction and inflamed areas around tonsillar pillars
5. Sicca syndrome of conjunctiva and mucous membranes
6. Drenching sweats often reported, but seen most frequently later in day

7. Raynaud's phenomenon with infrequent loss of normal fingerprint
8. Unequal pupils and contrary pupil reaction to light
9. Tongue tremor
10. Rare Adie's pupil with absent patella reflex
11. Positive modified Romberg
12. Frequent equivocal Babinski/plantar reflex on one side
13. Cogwheel leg raising and lowering motion that increases during the day
14. Frequently reported muscle twitching; infrequently seen in office after exercise
15. Sometimes marked falling pulse pressure in arterial pressures taken first when prone, then sitting, then standing
16. Rapid heart rate on minor activity such as standing
17. Associated fibromyalgia
18. Unusual sensitivity of cervical vertebrae area
19. Laryngeal stridor when fatigued
20. Nodular thyroid

EXAMINING TEST RESULTS

The patient may have had a large number of tests and physician reports. These should be examined in detail. Sometimes these tests disclose the clues to diagnosis that have been missed. Never simply accept an **MRI**, **PET** (positron emission tomography), **SPECT**, or X-ray report that states it is normal. When possible, review the film or printout yourself. If you feel uncomfortable doing that, find a specialist who can assist you. Recently, I saw a patient who had been seen at a major U.S. neurological clinic in Boston for three days at a cost of $12,000. Part of this examination included an **MRI** that was read as normal. When I asked a neuroradiologist to check this for me, he stated there was a significant lateral shift of the ventricles and to look for a malignancy or atrophic condition.

Some M.E./CFS patients (e.g., patients with spherocytosis and sickle cell anemia), tend to have an unusually low erythrocyte sedimentation rate (ESR). Elevation of ESR may suggest an active inflammatory disease. Persistent elevation may indicate an acute infection, a malignancy, or a missed rheumatoid disease. ESR is an inexpensive nonspecific test, and some medical organizations impugn it. All high ESR patients should be rechecked for chronic infectious, malignant, and rheumatoid disease. Physicians should always repeat all abnormal tests before coming to any conclusions since it may be a false abnormal.

ORDERING TESTS

Routine Blood and Chemical Tests I ask patients to have their own physician do locally any tests that have not already been done. It saves patients both time and money. Most physicians are happy to assist, but a few simply refuse, perhaps because they believe that the patient has no measurable physical illness. For example, glucose tolerance tests are increasingly frowned on by significant members of the medical community as being expensive and unnecessary in the evaluation of diabetes mellitus. These physicians are correct; however, a glucose tolerance test does more than simply define diabetes – it can demonstrate *hypoglycemia,* that much maligned illness that has passed out of vogue. I always do insulin levels with my glucose tolerance tests and frequently discover derangement of insulin response in some M.E./CFS patients.

The following tests should be considered for all M.E./CFS patients:

1. Routine CBC with sedimentation, blood smear, ferritin, and IBC. Many patients have a significant ferritin and IBC anomaly with normal Hb and Ht.
2. Eosinophil count.
3. Before ordering B12, check with the patient, who often is consuming vast amounts of B12 in vitamin combinations that will give abnormal highs.

4. Urinalysis and culture.
5. Immune and protein electrophoresis.
6. Immune panel only if it can be done in the immediate vicinity.
7. TSH, FT3, FT4, and thyroid antibody tests.
8. Thyroid ultrasound must be done on all patients. In the past two years, I have diagnosed six cases of thyroid malignancy with ultrasound. Often, these patients have normal serum thyroid tests.
9. Parathyroid Ab, Ca, and Ph.
10. Complete lipid profile.
11. HIV 1 and 2, treponema antibodies, hepatitis B (surface and core ab) and C, toxoplasmosis, histoplasmosis, Lyme disease.
12. Tuberculin skin test for all patients who have not received immunization.
13. Stools for parasites, ova, and blood x 3.
14. SGOT (AST), SGPT (ALT), bilirubin, BUN, uric acid.
15. ANA and rheumatoid battery if suggested.
16. PA and lateral X-ray of chest and X-ray of sella tursica and sinuses.
17. Fasting and 3-hour glucose and glucose tolerance if indicated.
18. Smooth and striated muscle ab and mitochondria ab.
19. Street drug profile to include cannabis, cocaine, LSD, and so on.
20. Prostate specific antigen (PSA) on all males over 25.

Thyroid Disease   It is well known that the thyroid is one of the essential glands that regulate energy and temperature, and it is equally well known that M.E. /CFS patients tend to have both energy and temperature disregulation. For this reason, I not only do free T4 and TSH on all patients but also do thyroid antibody tests. Even with major thyroid disease, the TSH may be normal. TSH appears to have a diurnal rhythm as do cortisol levels; TSH may vary from week to week. Even with all of these tests returning as normal, I do a thyroid ultrasound on all M.E./CFS patients. I then do a needle biopsy on all hypervascular nodules found or solitary nodules over 1 cm in diameter. Thyroid ultrasound is noninvasive and inexpensive. I examined one patient who had been seen by over 20 physicians in the United States and found a malignant thyroid. In the past 18 months, I have discovered 5 M.E./CFS patients with a malignant thyroid requiring thyroidectomy and three with missed Hashimoto's thyroiditis. Curiously, each of these patients with a malignant thyroid also had a history of spending much of the day before a computer terminal at work. I do not know if the computer terminal association is more than a simple fortuitous association. These patients with significant thyroid pathology as found on ultrasound and biopsy often have relatively normal TSH, free T4 and, less frequently, relatively normal thyroid antibody tests. Their thyroid pathology, however, is only part of a general autoimmune dysfunction, certainly involving the CNS but undoubtedly other areas as well. NeuroSPECT scans in these patients, as well as their immune tests also tend to be grossly abnormal. Once the thyroid problem is successfully treated, the patient occasionally gets better, but more often does not. The SPECT immune anomalies tend to persist.

Discussion: For some, this list of tests would already appear to be excessive. However, I cannot count the times that I have found abnormal thyroid and parathyroid function in this group of patients. HIV does not normally cause a fatigue syndrome except in its final stages and I rarely find HIV or positive treponema tests in this group. I may find one HIV every two years, but every year I discover several unexpected cases of either hepatitis B or C in M.E./CFS patients. Some of these patients are sexually active and have seen 20 physicians who have not ordered these tests or discovered these illnesses.

Low levels of elevated ANA are almost to be expected in many M.E./CFS patients, particularly early in their illness. Over time, the ANA levels tend to fall in those who do not go on to develop clinical rheumatoid disease. I occasionally find scleroderma antibodies in patients with clearly defined Sicca syndrome. At least 50% of my patients have either significantly abnormal immune or, much less frequently, abnormal protein electrophoresis that sometimes leads to the diagnosis of specific diseases.

If immune tests are not done on the same day the blood sample is taken, the levels will not be correct. These tests are very expensive and serve little purpose for either patient or physician unless they are part of a total investigation and are performed in an expert laboratory. Immune abnormalities should be repeated with a suitable hiatus to make sure they do not reflect an acute infectious anomaly. I almost never find evidence of street drugs in these patients. However, it is amazing how few patients have had a chest X-ray in the previous decade, and at times I find major lung, mediastinum, and cardiac pathology. Although gross observable pituitary anomalies on routine X-ray occur with increased frequency in this group, they tend to be few and far between. One should be aware in doing pituitary or adrenal tests that birth control medication can cause variables in this group of tests that at first would appear pathological. I recently examined a beautiful teenage girl with a moon-shaped face, hair changes, and marked striae who had been treated with corticosteroids by her physician for adrenal dysfunction. She had all the usual corticosteroid side effects and all that was behind it was that she was on birth control pills. There was no adrenal dysfunction.

I have now seen several children previously diagnosed with CFS who proved simply to have intestinal parasites. Once treated, these patients immediately bounced back to health. A positive eosinophil count is a good indication to look for parasites.

I do a PSA on every male above 25. I had a very good friend, a wonderful cardiologist, who in a jovial manner told me one day he had CFS. I asked if I could examine him and he declined saying it would get better. His CFS was discovered, too late, to be a prostatic carcinoma. If there is a family history of prostate disease, I do a transrectal ultrasound as well as a PSA, which may pick up prostate malignancy later than one would want to.

Most physicians would not find the preceding series of tests all that alarming unless they believe that the M.E./CFS group of illnesses is an invented phenomenon. This simple set of tests may lead to other tests that define the disease of at least 25% of the group who are mistakenly diagnosed with M.E./CFS. Without this baseline, it is pointless to do more expensive tests since the tests already mentioned may suggest the illness. Even physicians who agree that this set of tests is reasonable may balk at additional tests.

TWO CAUTIONS

Testing for Legal Purposes   Although I have seen patients win disability claims with one good proof, it is better to be prepared to give at least three significant pieces of evidence demonstrating proof of disability. Then if one proof is discounted, your multiple evidence may be sufficient to win over the judge. Some investigations may take up to two years. If your patient's disability case will possibly go to court before then, you should urge the patient to first obtain a lawyer experienced in insurance law to advise both the patient and yourself. A lawyer will make sure that the insurance company does not invalidate the claim by creating delays. Some states in the United States and all insurance companies have widely different approaches to disability claimants. Your patient requires necessary legal protection from the onset. Do not delay this essential step of making sure your patient has a lawyer before you spend significant time on investigation. It does patients no good to find that they are chronically ill with little chance of recovery if they then also lose the right to claim on their disability insurance. An expert lawyer may be as essential to the welfare of these patients as an expert investigation.

Lumbar Puncture: There are two important points to be aware of in doing a lumbar puncture. The most important is that during the early days or weeks of the disease, the patient may have a significant increase in intracranial pressure. *Always use a small-bore needle.* Do not forget to take the spinal fluid pressure reading as well as the fluid.

Since 1985, I have seen two patients where the physician's use of a large-bore needle for the puncture resulted in the brain stem being herniated into the upper spinal canal causing a permanent iatrogenic partial paralysis. The second point to remember is that many patients with acute onset M.E. /CFS may demonstrate IgG oligoclonal bands in their spinal fluid. These do not usually go on to develop multiple sclerosis (MS). Do not frighten or advise patients prematurely that they have MS without meeting the full obligatory MS criteria.

VIRAL CAUSES OF ACUTE ONSET M.E. OR CFS

Thousands of physicians in North America, Europe, and Austral-Asia have expended considerable funds to study the possible viral causes of M.E. /CFS. Some physicians have their pet theories, but none has been proven to be correct. Viral antibody tests are a particular waste of time and money since all humans are a virtual bank of hundreds if not thousands of viruses, some of whose antibodies are reactivated with a wide range of viral challenges. Only in epidemic situations where a rising viral antibody titer can be captured is it worthwhile to do antibody tests. Since acute onset disease has an incubation period of 4 to 7 days, usually at the lower limit, it makes little sense to ruminate about herpes virus 6 with an incubation of approximately 10 to 12 days or EBV with an incubation period of around 40 days. Even with a positive SPECT as a marker, we have not found a consistent viral cause. But having said this, I should add that we have never found any chronic viral infection by PCR on any patient with definite gradual onset CFS.

In acute onset M.E. patients with clear SPECT changes, however, we have had positive enterovirus PCR in about 10% of these patients for up to 3 years post-illness onset. The enterovirus that we have found has often been a new, nonlisted enterovirus similar to ECHO 25. This positive finding begins to drop at 2 years, and we have not found an elevation after 3 years. We have not found this virus in normal healthy controls with the exception of two normal patients who had received massive blood transfusions. Another curious feature is that many acute onset M.E. patients have incredibly high polio 1, 2, or 3 antibody levels. They obviously do not have polio, but perhaps some of the viruses that cause acute onset M.E. are similar in nature to poliovirus. Even so, 10% of acute onset illness represents about 5% of the total number of M.E./CFS cases. If it is an answer, it is only a partial answer at best. Enterovirus PCR is also very difficult to perform and many North American labs do not have the experience to perform this test accurately. I noticed that a physician for the National Institutes of Health (NIH) was doing their enterovirus analysis in Scotland, where I was doing mine. The difference is that the NIH physician was not screening patients for acute and gradual on-set illness or investigating them in any detail to remove patients with major missed disease.

DOPPLER ULTRASOUND AND ECHOCARDIOGRAMS

The most important tests that I do are Doppler scans and echocardiograms. They are more productive than MRIs or almost any other group of tests in uncovering pathology in M.E./CFS patients. The following tests, which I do on all patients, pick up another 25% of the underlying cause of disease:

- Visual carotid Doppler from aortic arch
- Visual transcranial Doppler to include vertebral and basilar arteries
- Thyroid ultrasound
- Echocardiogram and Doppler

Discussion: Dr. John Richardson from Newcastle upon Tyne has followed M.E. patients in Durham and Northumberland counties of the United Kingdom for three to four generations. I am aware of no other physician in the world with such a historic view of M.E. patients. He has repeatedly demonstrated that many M.E. patients go on to develop structural heart injury. The injury is usually valvular or related to pericardial effusion, and although most settle down, some do not and may develop myopathy. So I started to look at the hearts of these patients.

I have found that during the first year of acute onset M.E. /CFS disability, the incidence of pericardial effusion is unusually high. This seems to settle down with no apparent short-term problem, and after a year, the cases of pericardial fluid decrease considerably. However, the incidence of valvular disease in people in their 30s and 40s appears to be higher than in the normal population. When I find a significant valvular injury, I then repeat the echocardiogram yearly, and more frequently if the patient develops shortness of breath. I have observed several cases of elevated right heart pressure, significant septal defects, and increased myocardial wall thickening. Some who have had the injured valve replaced have miraculously returned to normal health. Are these incidental findings? I do not know, but Dr. Richardson has identified more than several hundred cardiopathies in his M.E. practice. I had two heart valve replacements in this group in the past year out of a total of 50 new patients.

Carotid and Transcranial Doppler Few physicians investigating M.E. /CFS employ the visual carotid and transcranial Doppler. This is a major error. It is a relatively inexpensive and totally safe procedure that does things no other type of test can do. On rare occasions, you will find aneurysms and subclavian steal anomaly with this test. Carotid atherosclerosis – sometimes substantial – is often found in patients with lipid dysfunction. This is a treatable condition and can be part of the cause of a CNS fatigue syndrome. You may say that any internist or cardiologist can pick up carotid pathology with a stethoscope, but few do who do not have an office Doppler. The carotid scan is also essential if you wish to do a transcranial Doppler

I examined a patient from the United States who had been diagnosed as having CFS in two major U.S. CFS clinics. She was given alternative medications and told to return in one year. She had complete obstruction of the vertebral basilar arteries and approximately 80% and 90% obstruction in either carotid. I was amazed that she was still alive. She was successfully operated on in Boston and her CFS has significantly improved.

The transcranial Doppler is not a perfect test. Patients with small foramen magnum space are difficult to visualize. But it will demonstrate high level internal carotid and other arterial obstruction that is beyond the normal range of a stethoscope. Only rarely do I get the chance to investigate posttraumatic mitral valve area (MVA) patients who develop an acute fatigue syndrome where personality or intellectual change has given rise to the diagnosis of CFS. In two of the past four such patients, I have demonstrated small subcortical arterial blowouts that had been missed by neurologists and that were possibly the cause of their pathophysiology.

In patients with M.E. /CFS, it is possible to demonstrate spasmodic disease of both major and smaller arteries with no typical evidence of migraine. This arterial pathology may be the end organ underlying cause of some M.E. patients' illness. Often MRIs and MRAs miss such arterial physiological pathology. Why? The technology of the MRA consists of a receiving computer revolving around the brain that may only give a picture of the maximum arterial diameter. In other words, what you see on the MRA is not reality but one view of reality. With the transcranial Doppler, the operator actually watches and films the kinetic movement of the arteries within the brain and can measure the velocity of the blood flow. Not only can you see these arteries move; if they are in spasm, you can observe this as well. Like M.E. /CFS muscles, M.E. brains are sometimes in significant pathological spasm. This knowledge may lead to more effective treatments of M.E. /CFS disease. Arterial spasm may account for some, but not all, of the SPECT changes that are routinely seen in M.E. patients.

I often find partial or complete vertebral or basilar artery obstruction. Frequently, I find left middle cerebral artery spasm or obstruction and, less frequently, frontal artery spasm in M.E. /CFS patients who do not report a migraine history. Left middle cerebral arterial field hypoperfusion is typical of M.E. .

ULTRASOUND

Consider using the following ultrasound scans:

1. Abdominal and pelvic organs and aorta
2. Prostatic ultrasound
3. Femoral and popliteal arteries in patients with leg pain

Discussion: Like many physicians, I have never been able to palpate a spleen except in the most extreme cases, such as you find in malaria. (We don't have much malaria in Canada.) Like most physicians, however, I can pick up an enlarged spleen with ultrasound. Early on in the M.E. /CFS disease, you will find a small number of enlarged spleens, but this becomes infrequent as the disease progresses past one year. Fatty infiltration of the liver is regularly seen and is usually dietary. I infrequently discover metastatic cancer (CA) in the liver. It is rare to uncover other major organ pathology to account for CFS. Organ pathology is more common in women, where too frequently we have found ovarian and pelvic tumors – some malignant (Billy Wilder's wife was diagnosed with CFS in 1989 and nothing was done for her condition; she subsequently died of ovarian cancer). Ultrasound is a fairly inexpensive noninvasive type of testing, and I do it on every patient. I routinely find pelvic pathology in as many as 30% of females. In the past three years, I have found only three pelvic malignancies – fortunately, the discoveries saved two patients' lives.

FURTHER EXAMINATION OF THE HEART AND CARDIOVASCULAR SYSTEM

The following tests are recommended:

1. 24-hour Holter monitor
2. Stress ECG or chemical stress test
3. Cardiac PET scan

4. Circulating red blood cell and serum volume

Discussion: I routinely use a Holter monitor on all patients. The cardiologist often reports these as normal. Do *not trust this report.* What the cardiologist or computer is basing the report on is the number of ischemic events. However, read the lowest heart rate at night, and note that it sometimes falls to the low 40s. Though this may be normal in an athlete, it is not in a sedentary M.E./CFS patient. For a patient who is not active all day long and has an average heart rate that flirts with 100 beats per minute or more, you know that this is not normal. These abnormal tests, however, are often reported as normal. We routinely pick up significant abnormal ischemic events. Similarly, as high as 10% of our patients have coronary artery disease. This is verified on stress test. So often do I find significant ischemic hearts in this group despite their young age that I now do stress tests only in the cardiology department.

Patients with M.E./CFS frequently cannot do exercise tests, and so I then do chemical testing as a second best. Several of our patients have reacted severely to the chemical test with excruciating pain. This is not true angina, and although the pain sometimes ceases as soon as the chemical is stopped and the antidote given, sometimes it persists for weeks after the procedure with no sign of coronary artery disease. I do not understand this phenomenon, but it is obviously vascular. The cardiologists state that this pain does not occur with the same frequency in non-M.E./CFS patients and now recognize it as a sign of pain or fibromyalgia associated with M.E./CFS.

The cardiologists routinely do cardiac PET scans on my patients with positive Holters and to date have only very rarely found ischemic muscle pathology. Dr. Peter Behan from Glasgow has demonstrated routine abnormal myocardial PET scans on his M.E. patients (Behan & Behan, 1988; Behan et al., 1985). Once again, I have not figured out why he can get these and our cardiac unit cannot. I do circulating blood volume on all patients. Dr. David Bell, a pediatrician in New York, was the first to demonstrate this useful test. It serves little purpose for most physicians, however, unless they test all patients with the same protocol and the nuclear medicine department has experience with this test. In our hospital, we find a wide variety of circulatory changes in relation to surface volume. I have some M.E. patients with a circulating red blood cell volume less than 50% of expected and a very large number with the range of 60% to 70%. What this test means is that blood is pooling somewhere in the body and that this blood is probably not available for the brain. In effect, there may be a reduced perfusion of oxygen and a reduced perfusion of insulin, growth factor, and other essential nutrients and chemical triggers in these patients.

When blood flow to the heart decreases sufficiently, the organism has an increased risk of death. Accordingly, the human body operates in part with pressoreceptors that protect and maintain heart blood supply. When blood flow decreases, pressoreceptors decrease blood flow to noncardiac organs and shunt blood to the heart to maintain life. This, of course, robs those areas of the body that are not essential for maintaining life and means the brain, muscles, and peripheral circulation are placed in physiological difficulty. This may cause much of the symptoms in M.E. /CFS patients. It probably suggests an intrinsic autonomic failure in these patients. We see SPECT changes in the subcortical brain responsible in part for maintaining reasonable autonomic function. I *repeat,* this test – circulating blood volume – is not useful except in labs with technicians who can do it correctly, in patients who follow a precise protocol, and in relationship to the complete assessment. Many patients want to run off and obtain one and then say, "This proves I have M.E. or CFS." Neither this test nor any single test proves the presence of disease.

DIAGNOSTIC TESTS OF MYALGIC ENCEPHALOMYELITIS

Consider the following tests:

1. SPECT
2. Xenon SPECT
3. PET

4. Neuropsychological Testing

Discussion: The primary diagnostic criterion for M.E. is acquired CNS change. We have excellent tools for measuring these physiological and neuropsychological CNS changes: SPECT, xenon SPECT, PET, and neuropsychological testing. CFS patients may not have any of these findings, particularly if their illness is due to some of the problems previously discussed.

Ever since Dr. Jay Goldstein asked Dr. Ishmael Mena, Nuclear Radiologist then at Harbor-UCLA Medical Center, to measure brain dysfunction in an M.E. /CFS patient's brain with his SPECT scan (Goldstein, 1992; Mena, 1991), it has been one of the most important tests for me in the evaluation of M.E.

I do not describe a patient as having M.E. unless there is an abnormal SPECT. If the SPECT is normal, I often repeat it along with xenon SPECT. If the brain scans remain normal, I conclude that it is unlikely to be M.E. I then refer to the patient as a CFS patient and search for other causes of the fatigue syndrome. Few people listened to Drs. Goldstein and Mena at the time. The problem was not with the physicians but that the CFS definition did not fit their discovery. Their discovery did precisely fit M.E. .

What is a SPECT? This is a computer-driven technology that demonstrates the microcirculation of the terminal arterioles in the brain and/or the function of areas of the individual brain cells. An M.E. patient has an abnormal brain SPECT. In this technology, there is a microcirculatory phase and a cellular phase. To my knowledge, the technology cannot distinguish whether the problem is in the brain cell or in the microcirculation, or both, unless a two-phase test is performed.

In some cases of MRI spectography of arm muscle of M.E. patients, it has been shown that because of an abnormal buildup of normal metabolites, the muscle cell actually shuts down to prevent cell death. This cell field shutdown is probably what is happening to the true M.E. patient's cell physiology in the brain. It probably explains in part the so-called brain fog and the dysfunction after the brain is stressed. It probably also explains muscle dysfunction. In legal cases, I also attempt to send the patient for xenon scan, which demonstrates the significant shutdown of the brain after exercise. I also send the patient for PET scans to obtain confirmatory changes in this completely different technology. Neither xenon SPECT nor PET is necessary except for research or legal cases, but both give a great deal of information about the pathophysiology of this disease to the knowledgeable physician. Xenon SPECT scans are almost impossible to find.

In visual terms, SPECT changes come down to two basic types of radical changes to brain physiology. The typical SPECT change in an M.E. patient is a decreased perfusion in the cortex in the area of the left middle cerebral artery and the branches leading to the posterior parietal lobes. This can also affect the anterior cerebral artery on the same side. Less commonly, this pathophysiology occurs on both sides involving both left and right middle cerebral arteries and anterior frontal arteries. Still less frequently, the findings are noted primarily on the right. To my knowledge, no one has published on whether there is an increased right brain abnormality in left-handed patients. Among other functions, the left middle cerebral artery covers the areas of the brain for visual and auditory recognition and interpretation. Decreased function of this CNS area creates a significant memory problem in that the patient has difficulty laying down new information and retrieving old information in the presence of added information (any external sensory stimuli/stress).

Often there are also significant changes in the subcortical regions, specifically in the brain stem, cerebellum, and basal ganglia area. Some authorities have identified fibromyalgia changes in the immediate subcortical areas; however, I have not verified this. Another finding that we frequently discover is sometimes referred to as a *vasculitis* pattern (the "itis" part of myalgic encephalomyelitis). This change is identical to what one finds in a patient with HIV dementia. The pattern is irregular and basic SPECT structures appear distorted. Patients with this vasculitis pattern are some of our most severely affected. Once again, the referring physician needs to actually see the scan and be able to read it since the neuroradiologist often only does a partial report. The brain is a big area and you have to be able to ask the neuroradiologist to check specific areas for anomalies. This is where your ability to read these scans is important. There are also problems with the SPECT equipment. Unless the neuroradiologist has a reasonably up-to-date SPECT scanner and appropriate columneter and associated software, the results may simply not be as good as they could be. SPECT scans are only helpful in physicians who understand them; then they can be essential in the diagnosis. They should always be done in conjunction with a carotid and transcranial Doppler to rule out obstructive arterial disease, which is not uncommon in these patients.

NEUROPSYCHOLOGICAL TESTING

This is a complex type of testing, and a physician should attempt to locate an experienced neuropsychologist without ties to the insurance industry. Most neuropsychologists today are employed by the insurance industry, and if they find too much pathology, I suspect that they are no longer engaged. Do not be fooled by a negative insurance-paid neuropsychological report; psychologists whose primary training is not neuropsychology prepare many of these reports. Discussion of this area would require an expert such as Sheila Bastien.

EXPENSIVE TESTS

The following tests require sophisticated, costly equipment:

1. Magnetic resonance imaging
2. Magnetic resonance angiography
3. MRI pectography
4. Tilt table
5. Sleep function

Magnetic Resonance Imaging (MRI)   An MRI is what every patient wants, but which few patients need. But the physician may require the added certainty of the information it can provide. The MRI does not necessarily demonstrate the disease pathophysiology of an M.E. patient, but it may demonstrate the cause of a CFS patient's illness (Hyde, Biddle, & McNamara, 1992). Many physicians simply do an MRI of a patient and almost nothing else. When it comes back as normal, the physician or the insurance company then tells such patients that nothing is wrong with their CNS and so they can go back to work or school. It is a mistaken belief that a brain dysfunction can be seen in an MRI evaluation. Generally an MRI is useful for one purpose, to demonstrate abnormal anatomy. Some neuroradiologists do not read MRIs very well, particularly if they read the scans and not the computer module. On the computer, you can blow up areas in doubt and also obtain different images. Nevertheless, MRI is changing. It would be useful if all MRIs had the computer software to print out the volume of the entire brain and its subdivisions individually - the ventricles and certain brain organs such as the cerebellum. This capability is technically possible and exists in some centers. The brain areas could be measured for later comparison to facilitate observing rate of atrophy or increase of ventricular volume. There are changes on some M.E./CFS brains but what I use the MRI for is simply to rule out malignancy, ventricular or pituitary changes, and brain atrophy not seen clearly on a CT scan or routine radiography. This is important since I infrequently discover both MS and non-MS demyelinating and calcified areas suggestive of a previous focal infection.

   I have a patient who had a profound sleep and memory disturbance diagnosed as M.E./CFS after returning from Africa. The instant I stopped talking to her to answer the telephone, she would fall asleep in my office. One day, I let her sleep from in the morning until 5 in the evening simply to see when she would wake up. She didn't. So I woke her up at 5 o'clock to tell her it was time to go home. I sat there all day doing paperwork and she didn't once awaken. She was negative for all of the usual sleep-inducing illnesses common in Africa and had a totally normal MRI. The diagnostic clue in her history was that the African town in which she had stayed was overrun by millions of rats and she had been bitten. I made a presumptive diagnosis of Weil's disease with a CNS infection involving the sleep centers.

She gradually recovered over a period of 2 to 3 years only to fall ill again with a stroke and renewed sleep dysfunction. The repeat MRI at that time demonstrated calcification of the substantia nigra area of the basal ganglia and this Von Economo-like disease affecting the basal ganglia probably contributed to her profound sleep problem. Just because an MRI or any test is normal does not mean that it will stay normal. The patient's visible symptoms are sometimes a better barometer that something is wrong than the best tests that a scientist can devise.

Magnetic Resonance Angiography (MRA) An MRA is simply an MRI scan that highlights the arterial blood supply. An MRA demonstrates arterial obstruction but does not show the arterial activity that can sometimes be seen on transcranial Doppler. I have not done many of these, but they have been useful in posttraumatic cases of M.E. /CFS (Hyde, Biddle, et al., 1992). I have found more pathology in the less expensive transcranial Dopplers than in MRAs.

MRI Spectography Because these capabilities are few and far between, I have not used them except in the United Kingdom, in Oxford. MRI spectography has demonstrated the abnormal buildup of metabolic by-products during normal activity of M.E. patients muscle cells (Hyde, Biddle, et al., 1992). When this occurs, the cell effectively shuts down to prevent cell destruction. This is seen graphically when Olympic runners' muscles seize, often just before the finishing line. This may be what is happening in the brain cells that tend to shut down when stressed by normal activity.

Tilt Table Examination I frequently find gross abnormalities in M.E. /CFS patients with this test. A circulating blood volume and a complete cardiac investigation should be done first. This is not a test to undertake lightly since the patient's heart sometimes stops and may have to be restarted. This test should only be done in major hospital centers in the presence of an appropriate physician where such emergency capabilities can be instituted. With this test, I have found significant pathology in about 10% of M.E. /CFS patients suggesting significant autonomic disease. There is, however, a wide discrepancy in the protocols of this type of testing, and so it is difficult to compare results. Some physicians strap the patient onto the table preventing movement that would induce circulation and others leave the patient relatively free allowing some movement-induced circulation.

Sleep Function Testing This is a useful and important test for all patients. Many M.E. /CFS patients have sleep dysfunction that tends to be worse in the first year or two of illness. Often involuntary movements and pain spasms provoke this dysfunction, and the patient simply cannot obtain the necessary sleep. The chronicity of the disease poses a possible danger in placing these patients on long-term analgesics or hypnotics because the patient rapidly becomes accustomed to them while the overall pain and movement-induced sleep dysfunction persists. However, it is sometimes possible to correct obstructive sleep apnea when it is found.

PROS AND CONS OF ELECTROENCEPHALOGRAMS (EEGs) AND QUANTITATIVE ELECTROENCEPHALOGRAMS (QEEGs)

In court, I am sometimes asked why I did not do an EEG. There are major problems with EEGs. Although they are sometimes positive during the first few weeks of illness, they are rarely performed at this time unless the patients complain of severe headache during epidemics. An EEG only records activity on the outer millimeter of the brain. Almost all M.E. patients tend to have involuntary movement disorders that are worse when the patient first falls ill and tend to decrease with time. Yet the neurologist reading an EEG almost never states an EEG IS abnormal unless the patient has an active seizure disorder or is brain dead. An average neurologist will spend 10 minutes or much less reading an EEG, whereas it would take a PhD student weeks to measure any abnormalities in a scientific manner. Nevertheless, there is much to measure in an EEG, and what would take the PhD student weeks to read will take a computer microseconds.

A QEEG is simply an EEG attached to a computer that contains appropriate software. A QEEG will immediately demonstrate tumors and brain activity or lack of it related to specific stimuli that are simply not possible to detect on a non-computer-driven EEG. Using QEEG technology operated by an expert physician, we have been able to demonstrate not only lack of normal activity in M.E. patients but migration of the normal activity centers from injured areas to different parts of the brain. We have also been able to demonstrate that there is considerable interference between the damaged center and the new center. The patient can dampen this interference when not fatigued, but as the stress continues, the interference-dampening capability of the brain fails and the patient goes into a memory and CNS dysfunction situation. Research would be beneficial in this area.

FUNDAMENTAL ADVICE

The following guidelines are useful for all clinicians:

1. When a patient presents with fatigue with no obvious cause, the physician is obliged to search for an occult malignancy, cardiac disease, other organ disease, *or* chronic infection.
2. When a patient presents with changes in mentation, the physician should think in terms of atherosclerotic disease, other arterial injuries, or changes involving the brain. Some of these are easily and inexpensively measured with a Doppler exam.
3. Common things are common. First check out common illnesses that can cause brain dysfunction and fatigue.

4. Never trust your instincts in making a diagnosis. Bizarre behavior does not necessarily mean the patient needs psychiatric treatment, nor does totally normal behavior mean the patient is free from significant injury or illness.
5. If you cannot measure and confirm the dysfunction with scientific or physical tests, you cannot be certain that your diagnosis is correct.
6. Listen to and watch the patient carefully, and take a detailed history. Trust a patient's complaints, but do not trust a patient's self-diagnosis; patients are often wrong or caught up in the latest fad diagnosis. You spent years studying medicine to learn how to make the correct diagnosis, and on the basis of that diagnosis, you may be able to help the patient.
7. Once you have made a diagnosis or have found an anomaly to explain an illness, remember to complete your investigation – several illnesses or medical anomalies may be contributing to the ill health of a single patient. The injury that you find may be part of a larger spectrum of illnesses, some of it treatable.
8. Patients who arrive at the office of a new physician and who have been completely investigated by many excellent physicians are sometimes dismissed as psychiatric or faking. It is in these patients that I find all of the pathology, and some of it is obvious. Rarely do physicians do more than a routine series of tests. The belief that **CFS** is a psychological illness is the error of our time.
9. If you discover any significant but modest cardiac valve or other cardiac disease on first examination, repeat the Doppler or any essential test in 6 months if the patient develops shortness of breath, otherwise yearly, as this may represent a progressive injury.

If you fail to find any cause of disease, do not assume that the patient has a psychiatric disease or a school or work avoidance. Sometimes either your knowledge or the necessary technology has not been in place to make the diagnosis. Sometimes the patient's symptoms preclude easy recognition of the illness at that date.

CASE STUDIES

Case 1:

A new patient, a teacher, came in carrying reams of computer-driven notes. He was 52 years old and had been ill for 18 months. I started by simply asking him his name; a two-hour tirade followed. I did not attempt to stop him although his words were quite irrational. He berated the more than 10 physicians and 4 hospitals he had attended during the previous 18 months. He stated that they had sometimes asked him to leave after a few minutes without so much as examining him. He told me, "They had the nerve to infer that I was a psychiatric patient." He had lost his job as a teacher due to his illness, and these physicians were not willing to help him obtain his disability payments. He said he slept all the time. He said he was still tired after sleeping. He said that he had lost his memory and couldn't teach. He said that he had not received a disability insurance check for over a year. Yet it was apparent to me, and probably to his other physicians, that this patient had a mental illness. His physicians had referred him to psychiatrists who diagnosed a psychosis and placed him on antipsychotic medication. He reacted badly to these medications. He insisted they were wrong; he had diagnosed himself as having chronic fatigue syndrome. He conformed to the CFS definition. He did not complain of his severe obesity. He did not complain of his strong odor, but possibly the smell and the unstoppable irrational babble explained why some of his physicians had asked him to leave their office. The patient was obviously psychotic.

Yet his presumptive and actual diagnosis that was later confirmed was fairly easy to make. During the 2 hours in my office in which I simply sat back and let him talk, hoping he would slow down, he excused himself to urinate 4 or 5 times. Although I did nothing other than to listen to him, he was very happy when he left because I had not only listened but also asked him to come back. As he left, I gave him a requisition for a fasting and 3-hour blood sugar and a few other simple tests.

In four days when he returned, I was able to tell him that he was severely diabetic and that he had an extreme hyperlipidemia. I referred him to the diabetic clinic the next day, where treatment was started immediately. In the two weeks that followed, I was able to demonstrate that he had had a recent myocardial infarct. Within two weeks of the start of his treatment for diabetes, his psychosis had totally disappeared; he talked rationally in a perfectly normal manner; and I was able to take a reasonable history. He was placed in the hands of cardiologists, dieticians, and exercise physiologists. A letter of the findings was sent to his insurer and within 30 days he had received his back disability benefits from the previous year.

The physicians who had seen this patient were undoubtedly competent professionals. They had trusted their instincts and they had made a correct diagnosis. He did have a psychosis, but they did not go far enough to diagnose a relatively easily treatable diabetic encephalopathy. Yet had I been satisfied with the diagnosis of diabetes and diabetic encephalopathy, I would have missed the hypertension, the severe hyperlipidemia, and the myocardial infarct. This example shows why all patients who present with a self-diagnosis of M.E. or CFS, or a physician referral of M.E./CFS, should be given a complete and structured history, as well as a physical and technological examination. Unusual behavior does not always indicate primary psychiatric disease. Often patients have multiple disease problems to explain their illness.

Case 2:

This patient phoned me from the United States to ask for an appointment. The young man had visited two major U.S. clinics and had been seen by over 20 physicians in the United States and Great Britain. They had diagnosed him as either having CFS or psychiatric problems. He was a brilliant professional with an extremely high salary who simply wished to get back to the work and life he loved. His story is typical of many patients I have seen. He had been ill with a significant upper respiratory tract infection (URTI) including a severe sore throat. For some unknown reason, his physician decided to give him a combined hepatitis B and A immunization at this time. Within a week after the injection, he was severely ill with intellectual and fatigue dysfunction. He soldiered on for several months mainly through the courtesy of his associates and then finally had to stop. One of his insurance companies refused to pay him. Despite seeing many physicians and going to two of the most important medical clinics in the United States, he was no further along. He brought all of the tests performed on him over the previous two years. It proved to have been a very incomplete investigation, but there were clues. There was a minor TSH discrepancy on one of his tests. On physical exam, there was not much in his thyroid but it was a bit irregular. I ordered antithyroid antibodies that came back incredibly elevated, and the initial ultrasound came back as a nodular thyroid. A thyroid uptake scan came back with a diagnosis of Graves' disease, but this did not fit what appeared to be a Hashimoto's thyroiditis. I referred him to an endocrinologist in his hometown and asked for a biopsy. It came back as malignant plus Hashimoto's thyroiditis. The thyroid was removed, and he was placed on treatment for metastasis as a precaution.

However, his neuro-SPECT demonstrated a significant vasculitis pattern. What probably happened is that the hepatitis B portion of the immunization paralyzed his normal immune response to the ongoing infection. The existing presumed viral infection then became chronic. I have seen this scenario in many cases of post-hepatitis B immunization injury. The companies producing hepatitis B immunization serum now clearly state in their brochures not to immunize when the patient is ill. When these immunizations first came out, however, they were touted as being the safest immunization ever invented and the suggestion was that they could be given without any negative consequence. Immunizations are powerful tools and should not be used in a patient with an acute ongoing infectious illness. This patient has other injuries that I will not go into, but he has classical M.E. with the abnormal vasculitis pattern scan. His M.E. illness is still active, but hopefully will settle down.

Case 3:

This patient also called me from the United States. She had been seen and diagnosed as CFS by several clinics and M.E. /CFS physicians. It seemed foolish for her to waste her time and money. She had significant brain dysfunction and overwhelming fatigue. She had CFS. Her initial investigation took the morning, and I was able to obtain a carotid and transcranial Doppler that afternoon. The exam revealed severe hyperlipidemia and an 80% obstruction of both internal carotids and a complete obstruction of the basilar artery feeding the brain. Her cause of CFS was obvious, but experienced physicians investigating M.E. /CFS and other neurologists and internists in the United States had missed it. The obstruction in one of the arteries was removed and she improved.

Case 4 Group:

This Canadian government employee fell ill at a party, along with several other family members. She had typical acute onset M.E. including significant CNS dysfunction and rapid exhaustion with physical or intellectual stressors. Her twin daughters also fell ill and had to stop school. A teenage neighbor across the street who was not at the party fell ill the same week; all had the same symptom picture. The next-door neighbor developed leukemia and the neighbor three doors down was diagnosed with Crohn's disease all in the same two weeks. Was it coincidence? Possibly. The girl across the street was ill for 6 months with severe M.E. /CFS and then recovered totally and not only went back to her university but got top marks in her class.

One of the identical twins recovered and returned to university classes with minor ongoing problems. The other identical twin attempted to return to school but failed miserably. Five years later, the mother and the one twin are still ill. The SPECT brain scans are typical of M.E. . Why did one twin recover? With PCR (polymerase chain reaction), we were able to find that the mother and the one twin still ill had had a persisting and consistent novel enterovirus for over three years. The twin who recovered had no enterovirus and her immune system had gotten rid of this infection. The ill mother and daughter both have persisting antibodies to their mitochondria.

Another group also in Ottawa fell ill the same week, again at a party; all were family members. One had the same persisting enterovirus that we had followed for three years; the others we were not able to test because government cutbacks had closed our facility. The one woman with the persisting virus was a physical education instructor and government employee and was now suffering from chronic severe CNS problems and rapid exhaustion. A vasculitis SPECT brain pattern has persisted since she was first ill. She had severe dysautonomia and peripheral nerve injuries.

Interesting enough, her uncle, who fell ill at the same party, had CNS problems and profound exhaustion only after playing hockey, a very active sport. He developed rapidly advancing coronary valve injury and this valve had to be replaced. He has been advised to cease playing hockey and has done so; he is now active in golf without any further problems.

Another family member at the same party developed acute psychotic disease and recovered. The enterovirus finding is interesting. We have never recovered this virus in a gradual onset CFS patient. We have never found it in any normal controls except for two individuals who had both received massive blood transfusions and had heart disease. We have found this virus in only about 10% of acute onset patients and have not been able to recover it in patients after 3 years of illness. It is hard not to believe that some M.E. is not viral related.

CONCLUSION

Thirty years ago when a patient presented to a hospital clinic with unexplained fatigue, any medical school physician would have told the students to search for an occult malignancy, cardiac or other organ disease, or chronic infection. The concept that there is an entity called chronic fatigue syndrome has totally altered that essential medical guideline. Patients are now being diagnosed with **CFS** as though it were a disease. It is not. It is a patchwork of symptoms that could mean anything. The original concepts of searching for occult disease are relevant to patients presenting today with **CFS**, M.E., and other fatiguing illnesses. Furthermore, because you do not find pathology does not mean there is none. From 1985 until 1988, I investigated M.E./CFS patients in a manner that I thought was exhaustive. I found disease then in only about 10% to 20% of patients. Were the remaining 80% to 90% suffering from somatization, psychiatric disease, or simply faking? No, the error of analysis was in my ability. After I met and received training from specialists including Dr. Charles Poser, a neurologist at Harvard, Dr. John Richardson in the United Kingdom, Dr. Jay Goldstein in Beverly Hills, and Dr. Ishmael Mena, at the University of California-Los Angeles, my ability to diagnose a M.E./CFS patient's disability rose 50% to 80%. Today, I usually find the underlying measurable cause of disease in 70% to 80% of the patients that I investigate. Are the remaining 20% suffering from somatization or psychiatric disease? I don't think so. I think I am simply missing the underlying diagnosis.

REFERENCES

Acheson, E. D. (1992). The clinical syndrome variously called benign myalgic encephalomyelitis, Iceland disease and epidemic neuromyasthenia. In B. M. Hyde, J. Goldstein, & P. Levine (Eds.), *The clinical and scientific basis of myalgic encephalomyelitis/chronic fatigue syndrome* (pp. 129-158). Ottawa, Ontario, Canada: Nightingale Research Foundation Press.

Anderson, J. S., & Ferrans, C. E. (1997). The quality of life of persons with chronic fatigue syndrome. *Journal of Nervous and Mental Diseases, 185*, 359-367.

Behan, P., & Behan, W. (1988). Postviral fatigue syndrome. CRC, *Critical Reviews in Neurobiology, 4*(2), 157-178.

Behan, P., Behan, W. M., & Bell, E. (1985). The post viral fatigue syndrome, an analysis of the findings in 50 cases. *Journal of Infection, 10,211-222.*

Crowley, N., Nelson, M., & Stovin, S. (1957). Epidemiological aspects of an outbreak of encephalomyelitis at the Royal Free Hospital, London, in the summer of 1955. *Journal of Hygiene, 55*, 102.

Daugherty, S. A., Henry, B. E., Peterson, D. L., Swarts, R. L., Bastien, S., Thomas, R. S., et al. (1991, January). Chronic fatigue syndrome in northern Nevada. *Reviews of Infectious Diseases, 13*(Suppl. 1), S39-S44.

Fukuda, K., Straus, S. E., Hickie, I., Sharpe, M. C., Dobbins, J. G., & Komaroff, A. (1994). The chronic fatigue syndrome: A comprehensive approach to its definition and study. *Annals of Internal Medicine, 121*, 953-959.

Gilliam, A. G. (1938). Epidemiological study of an epidemic, diagnosed as poliomyelitis, occurring among the personnel of the Los Angeles County General Hospital during the summer of 1934. *Public Health Bulletin, 240.*

Goldstein, J. A. (1992). Chronic fatigue syndrome: Limbic encephalopathy in a dysfunctional neuroimrnune network. In B. M. Hyde, J. Goldstein, & P. Levine (Eds.), *The clinical and scientific basis of myalgic encephalomyelitis/chronic fatigue syndrome* (pp. 400-406). Ottawa, Ontario, Canada: Nightingale Research Foundation Press.

Grufferman, S. (1992). Epidemiologic and immunologic findings in clusters of chronic fatigue syndrome. In B. M. Hyde, J. Goldstein, & P. Levine (Eds.), *The clinical and scientific basis of myalgic encephalomyelitis/chronic fatigue syndrome* (pp. 189-195). Ottawa, Ontario, Canada: Nightingale Research Foundation Press.

Henderson, D. A., & Shelokov, A. (1959a). Epidemic neuromyasthenia-clinical syndrome? *New England Journal of Medicine, 260*(15), 757-764.

Henderson, D. A., & Shelokov, A. (1959b). Epidemic neuromyasthenia-clinical syndrome? (concluded). *New England Journal of Medicine, 260*(16), 814-818.

Henderson, D. A., & Shelokov, A. (1992). Epidemic neuromyasthenia-clinical syndrome. In B. M. Hyde, J. Goldstein, & P. Levine (Eds.), *The clinical and* scientific basis of myalgic encephalomyelitis/chronic fatigue syndrome (pp. 159-175). Ottawa, Ontario, Canada: Nightingale Research Foundation Press.

Holmes, G. P., Kaplan, J. E., Gantz, N. M., Komaroff, A. L., Schonberger, L. B., Straus, S. E., et al. (1988). Chronic fatigue syndrome: A working case definition. *Annals of Internal Medicine, 108, 387-389.*

Holmes, G. P., Kaplan, J. E., Schonberger, L. B., Straus, S. E., Zegans, L. S., Gantz, N. M., et al. (1988). Definition of chronic fatigue syndrome [Letter to the editor]. *Annals of Internal Medicine, 109,* 512.

Hyde, B. M. (1992). A bibliography of ME/CFS epidemics. In B. M. Hyde, J. Goldstein, & P. Levine (Eds.), *The clinical and scientific basis of myalgic encephalomyelitis/chronic fatigue syndrome* (pp. 176-186). Ottawa, Ontario, Canada: Nightingale Research Foundation Press.

Hyde, B. M., Biddle, R., & McNamara, T. (1992). Magnetic resonance in the diagnosis of ME/CFS, a review. In B. M. Hyde, J. Goldstein, & P. Levine (Eds.), *The clinical and scientific basis of myalgic encephalomyelitis/chronic fatigue syndrome* (pp. 425-431). Ottawa, Ontario, Canada: Nightingale Research Foundation Press.

Hyde, B. M., Goldstein, J., & Levine, P. (Eds.). (1992). *The clinical and scientific basis of myalgic encephalomyelitis/chronic fatigue syndrome.* Ottawa, Ontario, Canada: Nightingale Research Foundation Press.

Joyce, J., Hotopf, M., & Wessely, S. (1997). The relationship of chronic fatigue and chronic - fatigue syndrome: A systematic review. *Quarterly Journal of Medicine, 90,* 223-233.

Lloyd, A. R., Hickie, I., Boughton, C. R., Spencer, O., & Wakefield, D. (1990). Prevalence of chronic fatigue syndrome in an Australian population. *Medical Journal of Australia, 153,* 522-528.

Marshall, G. S. (1999). Report of a workshop on the epidemiology, natural history, and pathogenesis of chronic fatigue syndrome in adolescents. *Journal of Pediatrics, 134*(4), 395-405.

Mena, I. (1991, May 18). *Study of cerebral perfusion by neuro-SPECT in patients with chronic fatigue syndrome.* Presented at Chronic Fatigue Syndrome: Current Theory and Treatment conference, Bel Air, CA.

Peterson, D. L., Strayer, D. R., Bastien, S., Henry, B., Ablashi, D. V., Breaux, E. J., et al. (1992). Clinical improvements obtained with ampligen in patients with severe chronic fatigue syndrome and associated encephalopathy. In B. M. Hyde, J. Goldstein, & P. Levine (Eds.), *The clinical and scientific basis of myalgic encephalomyelitis/chronic fatigue syndrome* (pp. 634-638). Ottawa, Ontario, Canada: Nightingale Research Foundation Press.

Ramsay, A. M. (1988). *Myalgic encephalomyelitis and postviral fatigue states* (2nd ed.). London: Gower Medical.

Richardson, J. (1992). ME, the epidemiological and clinical observations of a rural practitioner. In B. M. Hyde, J. Goldstein, & P. Levine (Eds.), *The clinical and* scientific basis of myalgic encephalomyelitis/chronic fatigue syndrome (pp. 85-92). Ottawa, Ontario, Canada: Nightingale Research Foundation Press.

Sharpe, M. C., Archard, L. C., Banatvala, J. E., Borysiewicz, L. K., Clare, A. W., David, A., et al. (1991). A report - chronic fatigue syndrome: Guidelines for research. *Journal of the Royal Society of Medicine, 84*,118-121.

Shelokov, A., Habel, K., Verder, E., & Welch, W. (1957). Epidemic neuromyasthenia: An outbreak of poliomyelitis-like illness in student nurses. *New England Journal of Medicine, 257,* 345.

Sigurdsson, B., Sigurjonsson, J., & Sigurdsson, J. (1950). Disease epidemic in Iceland simulating poliomyelitis. *American Journal of Hygiene, 52,* 222.

Taylor, R. R., Friedberg, F., & Jason, L. A. (2001). *A clinician's guide to controversial illnesses: Chronic fatigue syndrome, fibromyalgia, and multiple chemical sensitivities.* Sarasota, FL: Professional Resource Press.

Wallis, A. L. (1955). *An investigation into an unusual disease seen in epidemic and sporadic form in a general practice in Cumberland in 1955 and subsequent years.* Unpublished doctoral thesis, University of Edinburgh, Scotland.

The Nightingale Definition of Myalgic Encephalomyelitis (M.E.)

Abstract

M.E. is a clearly defined disease process. CFS by definition has always been a syndrome.

It essential to define clearly Myalgic Encephalomyelitis. That is what the Nightingale definition of M.E. sets out to do. The definition is based upon two criteria: the excellent scientific work of respected physicians and scientists who investigated the various M.E. epidemics, and our modern scientific testing techniques and the knowledge resulting from examining thousands of M.E. patients using these techniques

Dedication

The following definition of Myalgic Encephalomyelitis (M.E.) was prepared as a result of an invitation to attend two meetings at the British House of Commons with the Honourable Dr. Ian Gibson, Member of Parliament for Norwich North. The first meeting was with Dr. Gibson and his parliamentary assistant, Huyen Le, on 27 October 2005.

The second meeting was with The United Kingdom Parliament Group on Scientific Research into Myalgic Encephalomyelitis (M.E.), composed of Members of the House of Commons and House of Lords. It was held at Portcullis House on 10 May 2006. The committee members included:

The House of Commons Committee on M.E.
- Dr. Ian Gibson (Labour MP for Norwich North)
- Dr. Richard Taylor (Independent MP for Wyre Forest)
- Rt Honourable Michael Meacher (Labour MP for Oldham West and Royton)
- David Taylor (Labour MP for North West Leicestershire)
- Dr. Des Turner (Labour MP for Brighton Hemptown)

The House of Lords Committee on M.E.
- Lord Leslie Arnold Turnberg (Labour) Royal College of Physicians
- Baroness Julia Frances Cumberlege (Conservative)
- The Countess of Mar

The Chairman of the joint committee, Dr. Ian Gibson, asked me to prepare a report that might assist the committee in its further deliberations. Here is what I recommended.

The Report

It became obvious to me that too much importance is being placed upon the definitions of Chronic Fatigue Syndrome, and not enough on the actual disease, Myalgic Encephalomyelitis. These two illness spectrums are not the same and should not be considered to be the same. Nor is there any doubt in my mind that the various definitions of CFS actively impede physicians' ability to make a rapid diagnosis and a scientific confirmation of the illness, thus preventing a possible immediate treatment of some of these significantly disabled M.E. patients.

The following definition and discussion, although completed after the tabling of the parliamentary report, has been nevertheless respectfully submitted to the Honourable Dr. Ian Gibson M.P. and his committee members of the House of Lords and Commons.

I hope that this definition will be helpful to Dr. Gibson and his committee in their deliberations and will give comfort to M.E. patients everywhere. It is a definition that allows physicians to diagnose and treat successfully some of these patients immediately. Many underlying pathologies of M.E. are already known, particularly the primary physiological vascular dysfunctions, but effective treatment is simply not available. This definition also suggests the direction that future research into these vascular pathophysiologies might take.

Preface

Since the Nightingale Research Foundation's publication in 1992 of the textbook, *The Clinical and Scientific Basis of Myalgic Encephalomyelitis / Chronic Fatigue Syndrome* (Hyde, B, 1992), there has been a tendency by some individuals and organizations to assume that M.E. and CFS are the same illness. Over the course of two International Association of Chronic Fatigue Syndrome (IACFS, formerly the American Association of CFS) conferences, there have been suggestions that the name CFS be changed to M.E., while retaining the CFS definitions (Holmes, G.P., Sharpe, M.C., Fukuda, K) as a basis for such change. This does not seem to me to be a useful initiative: it would simply add credence to the mistaken assumption that M.E. and CFS represent the same disease processes. They do not.

M.E. is a clearly defined disease process. CFS by definition has always been a syndrome.

At one of the meetings held to determine the 1994 U.S. Center for Disease Control (CDC) definition of CFS, in response to my question from the floor, Dr. Keiji Fukuda stated that numerous M.E. epidemics - she cited the Los Angeles County Hospital epidemic of 1934 (Gilliam, A.G.), the Akureyri outbreak of 1947-48 (Sigurdsson, B.) and the 1955-58 Royal Free Hospitals epidemics (Ramsay, A.M.) were definitely not CFS epidemics. Dr. Fukuda was correct.

The Psychiatric Label:

Unfortunately many physicians and some senior persons in governments, including Great Britain, Norway and to a lesser degree the USA and Canada treat CFS as a psychiatric illness. This view has been arrived at by some physicians' interpretations of the CFS definitions from the Center of Disease Control (CDC). Indeed, despite clear signals in the 1994 CDC definition that CFS is not a psychiatric disease (Fukuda, K.), each of the CDC definitions and their addenda referring to CFS remain open to interpretation as a psychiatric rather than a physical illness. This is not a view to which I subscribe. It is the CFS definitions themselves that give rise to this inaccuracy. Consider the following:

What other physical disease definitions essentially state that if you discover the patient has any physical injury or disease, then the patient does not have the illness CFS? In other words if you have CFS then it does not result in or cause any major illness. What else could CFS then be but any number of various psychiatric, social, hysterical or mendacious phenomena?

The various CDC administrations dealing with the subject have clearly stated that CFS is a physical, not a psychiatric disease. However, is there any other definition of any physical disease that is not provable by scientific and clinical tests? Only psychiatric diseases are not clearly verifiable by physical and technological tests.

What other physical disease definition requires a 6-month waiting period before the illness can be diagnosed? Any physician knows that to treat a disease adequately you have to be able to define the disease at its onset and treat it immediately in order to prevent chronic complications from arising. To my knowledge, in the entire history of medicine, there are simply no other disease definitions that have ever been assembled with a structure similar to the CFS definitions.

If you are still not convinced, check the Internet for the definition of: *DSMIII Somatization Disorder.* (DSM) You will find that there is little substantial difference to distinguish the DSMIII definition from the 1988 and 1994 CDC definitions of CFS. It is difficult to believe that the CDC medical bureaucracy is not aware of this similarity. It is thus understandable why the insurance industry, as well as some psychiatrists and physicians, have simply concluded that CFS, if it exists, is a somatization disorder.

I believe it essential to define clearly Myalgic Encephalomyelitis, returning the definition to its clinical and historic roots and complementing this information with the certitude of modern scientific testing. That is what the Nightingale definition of M.E. sets out to do. But let me first ask you a very important question.

What is the purpose of any medical definition?

What is the purpose of any disease definition if it is not to allow the physician to rapidly and accurately diagnose a specific illness in order to attempt to effectively treat the patient before the illness becomes chronic or to call in the appropriate specialists? Our definition solves this problem.

What then is the purpose of any disease definition, once the disease has become chronic, if it is not: (a) to elicit clues for the immediate effective treatment of at least some of the patients, (b) to separate out illnesses with a similar symptom pictures in order to effectively treat them and finally (c) to direct research into reversing pathophysiological injuries that can be defined in terms of modern testing but for which, there is no effective treatment. Our definition solves this problem.

There is a third purpose for any disease definition. That is to clearly define the spectrum and limits of the disease so that various physicians and researchers can clearly understand that they are talking about the same illness spectrum and so launch research into what will become an effective treatment. Our definition gives a clear baseline for investigation.

The Nightingale definition is based upon the following two criteria:

(a) The excellent scientific and clinical work of respected physicians and scientists who investigated the various M.E. epidemics.

(b) The results of modern scientific testing techniques and the knowledge accruing from examining thousands of M.E. patients using these and more historical techniques.

The proposed M.E. definition is designed to improve early diagnosis and treatment for the tens of thousands of patients stricken with M.E. It is not a new definition of CFS nor should it be conceived as a rewording of any previous CFS definition. What follows is the primary M.E. definition for adults.

The Definition

Primary M.E is a chronic disabling, acute onset biphasic epidemic or endemic (biphasic) infectious disease process affecting both children and adults. There are both central and peripheral aspects to this illness.

The Central Nervous System (CNS) symptoms, as well as the clinical and technological abnormalities, are caused by a diffuse and measurable injury to the vascular system of the Central Nervous System. These changes in the organization of the CNS are caused by a combined infectious and immunological injury and their resulting effect on CNS metabolism and control mechanisms. Much of the variability observed in an M.E. patient's illness is due to the degree and extent of the CNS injury and the ability of the patient to recover from these injuries.

A significant number of the initial and long-term peripheral or body symptoms, as well as clinical and technological body abnormalities in the M.E. patient, are caused by variable changes in the peripheral and CNS vascular system. The vascular system is perhaps the largest of the body's organs and both its normal and patho-physiological functions are in direct relationship to CNS and peripheral vascular health or injury, to CNS control mechanisms and to the difficulty of the peripheral vascular system and organs to respond to CNS neuro-endocrine and other chemical and neurological stimuli in a predictable homeostatic fashion.

C) When pain syndromes associated with M.E. occur, they are due to a combined injury of (i) the posterior spinal cord and / or posterior root ganglia and appendages, (ii) patho-physiological peripheral vascular changes, and (iii) CNS pain reception homeostasis mechanisms.

Depending upon the degree and extent of the ongoing CNS and peripheral vascular injuries, these patho-physiological changes in turn may give rise to both transient and in many cases permanent systemic organ changes in the patient.

As with any illness, the diagnostic criteria of M.E. are divided into two sections:

(a) The clinical features and history of the ill patient that alert the physician to the initial diagnosis, and

(b) The technological examinations that confirm to the physician proof of his diagnosis.

Clinical Features

The clinical features of Myalgic Encephalomyelitis are consistent with the following characteristics that can easily be documented by the physician.

1. **M.E. is an acute onset biphasic epidemic or endemic (sporadic) infectious disease process:** Both Epidemic and Non-Epidemic cases are often preceded by a series of repeated minor infections in a previously well patient that would suggest either a vulnerable immune system, or an immune system subject to overwhelming stressors such as: **(a)** repetitive contact with a large number of infectious persons, **(b)** unusually long hours of exhausting physical and / or intellectual work, **(c)** physical traumas, **(d)** immediate past immunizations, particularly if given when the patient has concurrent allergic or autoimmune or infectious disease or if the patient is leaving for a third world country within three weeks of receiving the immunization, **(e)** epidemic disease cases whose onset and periodicity appear to occur cyclically in a susceptible population, **(f)** the effect of travel, as in exposure to a new subset of virulent infections, or **(g)** the effects of starvation diets. (It should be noted that subsets c, d, e, f and g are all stressors associated with decreased immune adaptability **plus** an associated infection with an appropriate neurovascular infectious virus or other infectious agent. This may be due either to an immediate preexisting infectious disease or to a closely following infection, either of which may or may not be

recognized.)

Primary M.E. is an acute onset biphasic epidemic or endemic (sporadic) infectious disease process, where there is always a measurable and persistent diffuse vascular injury of the CNS in both the acute and chronic phases. Primary M.E. is associated with immune and other pathologies.

1. **Primary Infection Phase**: The first phase is an epidemic or endemic (sporadic) infectious disease generally with an incubation period of 4 to 7 days; in most, but not all cases, an infection or infectious process is evident. (See *Clinical and Scientific Basis of M.E./CFS, Chapter 13, pps. 124-126*)

1. **Secondary Chronic Phase**: The second and chronic phase follows closely on the first phase, usually within two to seven days; it is characterized by a measurable diffuse change in the function of the Central Nervous System. This second phase is the persisting disease that most characterizes M.E.

1. **The Presence or Absence of Various Pain Syndromes is highly variable**: The pain syndromes associated with the acute and chronic phases of M.E. may be described as **Early** and **Late** findings. **Early Findings:** (a) severe headaches of a type never previously experienced; (b) these are often associated with neck rigidity and occipital pain; (c) retro-orbital eye pain; (d) migratory muscle and arthralgia pain; (e) cutaneous hypersensitivity. **Late Findings:** Any of the early findings plus (f) fibromyalgia-like pain syndromes. This is only a partial list of the multiple pain syndromes. Many of the pain features tend to decrease over time but can be activated or increased by a wide range of external & chemical stressors. (See *Clinical and Scientific Basis of M.E./CFS, Chapter 5, pps. 58-62*)

Testable & Non-testable Criteria

The technological tests listed below can be used to (a) confirm the clinical diagnosis of Myalgic Encephalomyelitis and (b) to some degree gauge its severity and probability of persistence. The second and chronic phase that clearly defines M.E. is characterized by various measurable and clinical dysfunctions of the cortical and/or sub-cortical brain structures.

1. **Diffuse Brain Injury Observed on Brain SPECT**: If the patient's illness is not measurable using a dedicated brain SPECT scan such as a Picker 3000 or equivalent, then the patient does not have M.E. For legal purposes these changes may be confirmed by PET brain scans with appropriate software and/or QEEG. These changes can be roughly characterized as to severity and probable chronicity using the following two scales: **A)** Extent of injury and **B)** degree of injury of CNS vascular function.

1. **Testable Neuropsychological Changes:** There are neuropsychological changes that are measurable and demonstrate short-term memory loss, cognitive dysfunctions, increased irritability, confusion, and perceptual difficulties. There is usually rapid decrease in these functions after any physical or mental activity. Neuropsychological changes must be measured in relation to estimates of prior achievement. This feature may improve over a period of years in patients with adequate financial and social support and can be made worse by chronic stressors.

The neurophysiological changes are those observed by a qualified Neuropsychologist with experience in examining this type of disease spectrum. Some of the deficits that a Neuropsychologist should consider examining include: (a) word finding problems, (b) Subtle problems with receptive and expressive aphasia, (c) Decreased concentration, (d) Distractibility and the decreased ability to process multiple factors simultaneously, (e) Dyscalculia, (f) Decreased fine and gross motor problems, (g) Dysfunction of spatial perception, (h) Abstract reasoning, (i) Compromised visual discrimination, (j) Sequencing problems. In Cochran's Q Neuropsychological tests there is an increased observation of significant problems in both immediate and delayed verbal recall. In Dr Sheila Bastien's investigations, over 50% of M.E. patients have delayed visual recall, TAP dominance, TPT N-Dominance and 40% or more have abnormalities of Immediate visual recall, Tap N-Dom, Grip N Dominance, & grip dominance problems *(Bastien, Sheila. The Clinical and Scientific Basis of M.E. / CFS. Chapter 51, pps. 453-460)*

Extent of Injury:

Type 1: One side of the cortex is involved. Those patients labeled as 1A have the best chance of recovery.

Type 2: Both sides of the cortex are involved. These patients have the least chance of spontaneous recovery.

Type 3: Both sides of the cortex, and either one or all of the following: posterior chamber organs, (the pons and cerebellum), limbic system, the subcortical and brainstem structures are involved. Type 3B are the most severely affected patients and the most likely tobe progressive or demonstrate little or no improvement with time.

Degree of injury:

Type A: Anatomical integrity is largely maintained in the Brain SPECT scan.

Type B: Anatomical integrity is not visible in the CNS SPECT scan. Type 3B are some of the most severely and chronically injured patients.

1. **Testable Major Sleep Dysfunction:** This can include all forms of sleep dysfunctions. All or any of the following may be present: (a) impaired sleep efficiency, (b) significant fragmented sleep architecture, (c) movement arousals, particularly if there is

an associated pain syndrome, (d) absence or significant decrease of type 3 and 4 sleep, (e) abnormal REM sleep pattern (f) changes in daytime alertness and (g) sleep reversals.

1. **Testable Muscle Dysfunction:** This feature may be due to vascular dysfunction or peripheral nervous or spinal dysfunction and includes both pain and rapid loss of strength of muscle function after moderate physical or mental activity. This feature tends to improve over a period of years but many patients frequently remain permanently vulnerable to new disease episodes. Few centres are equipped or funded to make these examinations. Unfortunately only a few major medical centres are equipped to study this type of dysfunction.

1. **Testable Vascular & Cardiac Dysfunction:** This is the most obvious set of dysfunctions when looked for and is probably the cause behind a significant number of the above complaints. All moderate to severe M.E. patients have one or more and at times multiple of the following vascular dysfunctions. As noted, the primary vascular change is seen in abnormal SPECT brain scans and clinically most evident in patients with:

a) POTS: severe postural orthostatic tachycardia syndrome. **Note:** This group can be confused with diabetes insipidus due to the fact that they may have polydipsia from their attempt to increase their circulating blood volume by consuming large amounts of fluids. This group can be verified by the absence of pituitary adenoma or pathology and the fact that they can sleep through the night without waking to drink fluids (Streeten, David.) Despite the great steps forward in the understanding of this relatively common pathophysiology seen routinely in M.E. patients, a pathology which is really related to either an autonomic injury to the CNS, injury to the vascular receptors or both, very little of the present treatment protocol is of much use. The situation is so bad that few major centres have any well-funded expertise in either autonomic or vascular receptor injury. Many of the M.E. patients that are dismissed by physicians as suffering from lack of activity have significant proprioceptive injuries in these areas. Nor can we always rely on the few autonomic laboratories and their tilt table testing abilities. Many of the tilt table examination reports return as normal, many as grossly abnormal. Yet all the physician has to do is have each M.E. patient stand for 8-12 minutes to realize that a large number of these normal tilt table patients simply cannot maintain a normal blood pressure and normal heart rate. Compare this to non-M.E. patients and one immediately can tell the difference. A large number of M.E. patients have significant autonomic difficulties.

b) Cardiac Irregularity: on minor positional changes or after minor physical activity, including inability of the heart to increase or decrease in speed and pump volume in response to increase or decrease in physical activity. (Hyde, B., Chapter on Cardiac Aspects): (Montague, T.,) Cardiac irregularity is closely related to the above discussion. In many M.E. patients there is an unusual daytime tachycardia, particularly since these patients are often very sedentary. In doing a 24-hour Holter monitor this may be missed since the 24 hour average is usually given. One should always ask for wake time and sleep time heart rates.

c) Raynaud's Phenomenon: vasoconstriction of small arteries or arterioles of extremities, with change in colour of the skin, pallor and cyanosis. It is associated with coldness and pain of extremities. This is in part, the cause for temperature and pain dysfunctions seen in M.E. This phenomenon is found in many other conditions than M.E. Some of the associations are post-traumatic, neurogenic conditions, occlusive arterial diseases, toxic chemical associations and a wide range of rheumatoid conditions. Many of these conditions have associations with M.E. (See Magallni, S. for more detail.)

d) Circulating Blood Volume Decrease: This is a nuclear medicine test in which the circulating red blood cell levels in some M.E. patients can fall to below 50%, preventing adequate oxygenation to the brain, gut and muscles. These patients do not generally have aenemia and are not blood deficient. This is undoubtedly a subcortical dysregulation. It is associated with serum and total blood volume measurements. This is a concept that many physicians have difficulty understanding. I have heard physicians repeatedly tell the patient they are not aenemic and therefore dismiss this important finding. **Note:** So where does the blood go? Body servomechanisms are genetically designed so that blood flow and oxygen to the heart are always protected. Thus, when the body of the M.E. patient is stressed, the blood flow to organs not necessary for short-term survival, such as the brain, the gut and skeletal muscles, can be temporarily decreased. This of course gives rise to many of the M.E. symptoms.

e) Bowel Dysfunction: vascular dysfunction may be the most significant causal basis of the multiple bowel dysfunctions occurring in M.E. (See d. above.)

f) Ehlers-Danlos Syndromes Group: This is a group of illnesses with a genetic predisposition to M.E. or M.E.- like illness. In fact it probably represents a spectrum of illnesses that start with (i) hyper-reflexia syndrome, moving through any of the (ii) various Ehlers-Danlos Syndromes and climaxing in (iii) Marfan Syndrome where there tends to be early death if the aortic and cardiac changes are not repaired. Ehlers-Danlos syndromes can go undetected until what appears to be a switch is turned on, usually in late teens to early thirties. The "switch" may be viral or possibly age or hormonal related. Raynaud's phenomenon is usually associated. **Diagnosis:** briefly, patients over the age of 16 who can (i) touch their nose with their tongue, (ii) touch their forearm with the thumb of the same extremity (joint laxity), (iii) touch the floor readily with the full palm should be considered suspect for further examination. There are several fascination variations of Ehlers-Danlos. They are generally considered to be a group of genetic illnesses but in my examination of M.E. patients most often are not manifested until well past puberty and in adulthood. Additional generalized features of this spectrum of illnesses include (v) India rubber or hyperelastic skin, (vi) easy bruisability (vascular fragility), (vii) Arachnodactyly (long spiderlike fingers). Many of the patients with a more severe form tend to be tall, slender with a dolichocephalic skull, high palate and long narrow feet with hammertoes verging on Marfan syndrome. (See Magalini, S. I., Magalini S. C. for both E-D Syndrome and Marfan 1 and Marfanoid hypermobility.)

g) Persantine Effect in M.E. Patients: Persantine is a chemical manufactured by Boehringer Ingelheim. It is employed to perform chemical cardiac stress testing when a patient cannot exercise sufficiently to stress the heart. It is a particularly safe medication but when employed with many M.E. patients it can cause severe muscle pain over the extremities and entire musculature. Normally this can be reversed by injection of an antidote but this does not always work rapidly in M.E. patients. Severe pain and fatigue can be intolerable and persist for minutes to days in some M.E. patients following Persantine use. Persantine works by dilating both peripheral and cardiac blood vessels and causing the heart rate to increase as in a POTS patient. Obviously one major pain and fatigue factor in M.E. patients is caused by abnormal dilatation of peripheral blood vessels. The resulting pain may be related to reflex vasospasm as in severe Raynaud's phenomenon that I note elsewhere is one of the causes of M.E. pain. To my knowledge, no testing of M.E. patients with Persantine has ever been published by Boehringer Ingelheim or others. It is one of the reasons I believe that pain syndromes in M.E. patients are due to a pathological vascular physiology.

h) M.E. Associated Clotting Defects: M.E. represents both a vasculitis and a central and peripheral change in vascular physiology. All such vascular illnesses should be potentially treatable. We do not yet know how to adequately treat the (i) genetic forms of vasculitis & vascular patho-physiology mentioned here, nor (ii) the probable viral triggered genetic vascular pathologies also mentioned. Nor do we know how to treat those (iii) centrally caused injuries causing the circulating blood volume defects that are demonstrated when we do the "nuclear medicine circulating blood volume tests. It is important to do this test on all patients. POTS is poorly treatable and more often success in treatment presently escapes physicians' ability. Eventually, I have no doubt that these will be treatable causes of M.E. type disease. However there is a significant group of M.E. patients who are ill due to a treatable form of vasculitis and can be treated if the physician takes the time to diagnose the subgroup. These patients are the clotting defect patients. Some of these clotting defects are genetic and some appear to be genetic with an age or viral switching mechanism, as I have mentioned elsewhere with Ehlers Danlos Syndromes; although they may develop in childhood, they are more frequently noted well after puberty and before the age of 40. Many of these patients can be diagnosed by the following tests: (1) Serum viscosity test, (2) Antiphospholipid Ab., (3) Protein C defects, (4) Protein S defects, (5) Factor V Leiden defect, to name the most common that we have uncovered. However, there are others for which we also test. These conditions are all potentially treatable and when treated adequately may allow the patient to return to school or work. Although any physician can order these tests, a haematologist should review all M.E. patients for these and other possible clotting anomalies. Most clotting defects are treatable and treatment has resulted in recovery in some cases. Remember M.E. is essentially a problem of microcirculation and any improvement in this area can have dramatically positive effects. It is well worthwhile for all physicians reading this definition who have an interest in M.E. to examine the Internet for Hughes Syndrome. Curiously, Hughes Syndrome was first outlined in St. Thomas' Hospital London, the home of the Nightingale School of Nursing. Hughes Syndrome, a vascular syndrome also called Sticky Blood Syndrome, closely parallels the definition of M.E.

i) Anti-smooth muscle Antibodies: This is an antibody to the muscle tissue in the arterial bed. It is elevated in about 5% of M.E. patients but whether this is different in non-M.E. patients is unknown but unlikely. It rarely is over 1:40.

j) Cardiac Dysfunction: There are a large number of cardiac dysfunctions that can regularly appear in an M.E. patient. Certain are obvious and discussed under **Ehlers-Danlos** Syndrome and Marfan syndrome. I also discussed cardiac dysfunction in Chapter 42, *The Clinical and Scientific Basis of M.E./CFS*. Since that chapter was written a large number of other cardiac pathologies and pathophysiologies have been noted by various researchers and clinicians, particularly by Dr. Paul Cheney. Without a clear understanding of these significant problem areas **it is simply indefensible and potentially dangerous to place an unsuspecting patient in a graduated exercise program.** This is particularly true if the patient is not being tested in a cardiac unit. Although in our clinic we have performed what we believe to be a complete cardiac assessment on all patients seen, what the Ottawa Cardiac Institute and I believed was a complete assessment may be wanting. Over the next year we will reassess these patients with a more detailed cardiac examination and report on it in these diagnostic criteria.

1. **Testable Endocrine Dysfunction:** These features are common and tend to be of late appearance. They are most obvious in:

a) Pituitary-Thyroid Axis: Changes in serum TSH, FT3, FT4, Microsomal Ab., PTH, calcium and phosphorous rarely occur until several years after illness onset. This anomaly can best be followed by serial ultrasounds of the thyroid gland, where a steady shrinking of the thyroid gland may occur in some M.E. patients with or without the development of non-serum positive Hashimoto's thyroiditis (a seeming contradiction in terms) and a significant increase in thyroid malignancy. In cases of thyroid wasting, serum positive changes tend to occur only after years and often not until the thyroid gland shrinks from the normal 13 to 21 cc. volume in an average adult female and 15-23 cc. volume in male patients to below a volume of 6 cc. (Mayo Clinic averages) (Rumack, Carol). The normal serum analysis of patients for thyroid dysfunction, TSH, FT4, microsomal antibodies etc., the golden rule of most physicians and endocrinologists, is simply not an adequate means of ascertaining thyroid dysfunction in most M.E. patients. Repeat thyroid ultrasound must be performed for all M.E. patients to observe the presence of dystrophic changes. It is also inadequate simply to accept the radiologist's report of a normal thyroid. The volume of each lobe and its homogenicity must be requested and documented. Radiologists simply report normal thyroids when in effect they are hypo and hyper-trophic. Although the Mayo Clinic averages cited above may be criticized they are as good as any in ascertaining normal thyroid size.

The following changes, while uncommon, may also be related to an M.E. disease process:

b) Pituitary-Adrenal Axis Changes: where changes and findings are infrequent.

c) Pituitary-Ovarian Axis Changes

d) Bladder Dysfunction Changes: This dysfunction occurs frequently in the early and in chronic disease in some people. In some instances this may be due to a form of diabetes insipidus, in other cases it is related to POTS-type illness where the patient is compensating for the inability to maintain vascular pressure by attempting to increase fluid volume. In other cases this may be due to interstitial cystitis or a form of polio-type-bladder particularly if the cause of the individual disease is an enterovirus. Dr. John Richardson also associated this finding with adrenal dysfunction that he measured.

Discusssion

To various degrees many if not all of the above historic findings have been observed and discussed by Doctors Alexander Gilliam, Bjorn Sigurdsson, Alberto Marinacci, Andrew Lachlan Wallis, A Melvin Ramsay (Elizabeth Dowsett), John Richardson, Elizabeth Bell, Alexis Shelokov, David C Poskanzer, W.H. Lyle, Sir E. Donald Acheson, Louis Leon-Sotomayor, J. Gordon Parish and many others. Some of these features have not been noted previously.

To various degrees the following physicians have also noted many of the above historical and the more recent investigational findings. They include alphabetically, Doctors Peter Behan, David Bell, Dedra Buchwald, Paul Cheney, Jay Goldstein, Seymour Grufferman, Byron Hyde, Anthony L Komaroff, Russell Lane, Ismael Mena, Harvey Moldofsky, James Mowbray, Daniel Peterson, Vance Spence and scores of others. I have examined patients with M.E. since the late 1970s but only in 1985 at the urging of Dr. Charles Poser of Beth Israel Hospital at Harvard and John Richardson in Newcastle-upon-Tyne did I take up the study of these unfortunate patients on a full time basis. The material in this definition is the cumulative result of my listening and interpreting the work of all of the above clinicians and my evaluation of over 3,000 M.E. and CFS patients since 1984.

The essential concept of the indepth medical evaluation that is the basis of my work on M.E. and CFS since 1995 was crystallized in my discussions in Seattle Washington State in 2002 with Dr. Leonard A. Jason, Patricia A. Fennell and Renee R. Taylor. This discussion was set down as *Chapter 3, The Complexities of Diagnosis* in their book, *The Handbook of Chronic Fatigue Syndrome*, 2003, John Wiley and Sons, Hoboken, New Jersey (See Jason, Leonard A.). I would also like to thank Elizabeth Dowsett and Jane Colby whose work with children in the UK as well as their advice has been instrumental in this definition. I must also thank each and every one of the members of John Richardson's Newcastle Research Group who have provided me with so much valuable information over the years and who have all supported my continued investigations of M.E. patients.

What is new and different about the Nightingale M.E. definition is the following:

A. A Testable Definition: The definition is set out in such a fashion as to enable the physician to make a bedside or office clinical diagnosis and then to scientifically test the hypothesis. This will allow the physician an early diagnostic understanding of this complex illness and a scientific and technological method to investigate and confirm the diagnosis. It is well known by all serious physicians that in order to assist any patient in a partial or full recovery the illness must be (a) prevented from occurring by either immunization or understanding and avoiding the causes, (b) or diagnosed and treated immediately following onset. The Nightingale Definition assists the physician both in diagnosis and early treatment.

B. A Vascular Pathophysiology. The subject of vascular pathology is not new. The fact of the children dying of a Parkinsonian-like vascular injury to the basal ganglia in Iceland during the Akureyri M.E. Epidemic is an obvious indication of the CNS vascular effects in M.E. Vasculitis has been well documented by Dr. E. Ryll in his description of the epidemic in the San Juan Mercy, Sacramento California Hospital in 1975. He described this M.E. epidemic as an epidemic vasculitis.

He was correct. In the late 1980s Drs. Jay Goldstein and Ismael Mena confirmed and proved this initial description by examining the changed brain microcirculation using brain SPECT imaging in M.E. patients. Following my 21 years of examining M.E. and CFS patients and 16 years of subjecting the M.E. and CFS patients to brain imaging techniques suggested by Goldstein and Mena, it has become obvious to me that we are dealing with both a vasculitis and a change in vascular physiology. Numerous other physicians have supported this finding. Dr. David Bell, who rediscovered the work of Dr. David Streeten and his book, *Orthostatic Disorders of the Circulation,* advanced this understanding of M.E. The work of Dr. Vance Spence and his colleagues in Scotland have started to nail this CNS-vascular relationship down even further with a series of major research papers. The recent interpretation of the cause of Multiple Sclerosis (MS), as an injury of the microvasculization causing the injury of the schwann cells that in turn causes the demyelination injuries of MS has been added to that of paralytic poliomyelitis as an essential vascular injury. Paralytic poliomyelitis was thought to be a primary injury to the anterior horn cells of the spinal cord but is now recognized as a vasculitis injuring the circulation to the anterior horn cells. Poliomyelitis is generally a non-progressive, specific site injury, although post-polio syndrome with demonstration of subcortical brain changes has challenged that belief. MS is a recurrent more fulminant physiological vascular injury. M.E. appears to be in this same family of diseases as paralytic polio and MS. M.E. is definitely less fulminant than MS but more generalized. M.E. is less fulminant but more generalized than poliomyelitis. This relationship of M.E.-like illness to poliomyelitis is not new and is of course the reason that Alexander Gilliam, in his analysis of the Los Angeles County General Hospital M.E. epidemic in 1934, called M.E. atypical poliomyelitis.

C. The Lack of Mention of Fatigue: M.E. is not CFS: Fatigue was never a major diagnostic criterion of M.E. Fatigue, loss of stamina, failure to recover rapidly following exposure to normal physical or intellectual stressors occur in most if not all progressive terminal diseases and in a very large number of chronic non-progressive or slowly progressive diseases. Fatigue and loss of stamina are simply indications that there is something wrong. They cannot be seriously measured, are generally subjective and do not assist us with the diagnosis of M.E. or CFS or for that matter any disease process.

D. Cause: It is obvious that all cases of epidemic M.E. and all primary M.E. are secondary to infectious / autoimmune phenomena. Many M.E. and M.E.-like patients' illness is complicated by multiple other causes, some of which occur unnoticed prior to the illness and some that occur due to the illness itself. This is why a complete technological investigation has to be made on each chronically ill M.E. or M.E.-like patient. Under epidemic and primary M.E. there is no consensus as to the viral or infectious cause. Much of this lack of consensus may be due in large part to separate acute onset from gradual onset patients in the M.E. and CFS groups of patients. Primary M.E. is always an acute onset illness. Doctors A. Gilliam, A. Melvin Ramsay and Elizabeth Dowsett (who assisted in much of his later work,) John Richardson of Newcastle-upon-Tyne, W.H. Lyle, Elizabeth Bell of Ruckhill Hospital James Mowbray of St Mary's and Peter Behan all believed that the majority of primary M.E. patients fell ill following exposure to an enterovirus. (Poliovirus, ECHO, Coxsackie and the numbered viruses are the significant viruses in this group, but there are other enteroviruses that exist that have been discovered in the past few decades that do not appear in any textbook that I have perused.) I share this belief that enteroviruses are a major cause. Unfortunately, it is very difficult to recover polio and enteroviruses from live patients. Dr. James Mowbray developed a test that demonstrated enterovirus infection in many M.E. patients but I do not believe he qualified his patients by acute or gradual onset type of illness. In my tests in Ruckhill Hospital in viral infection only in acute onset patients and not in any gradual onset patients. Few physicians realize that almost all cases of poliovirus recovered from poliomyelitis victims came from cadavers. At the very least, these enteroviruses must be recovered from patients during their onset illness and this has rarely been done. An exception is in the case of the Newton-le-Willows Lancashire epidemic where Dr. W. H. Lyle's investigation recovered ECHO enterovirus. Recent publications by Dr. J. R. Kerr have also identified the fact that enteroviruses are one of the most likely causes of M.E. If this belief is correct, many if not most of the M.E. illnesses could be vanquished by simply adding essential enteroviral genetic material from these enteroviruses to complement polio immunization.

Non-Infectious M.E. Type Disease: I have not discussed noninfectious M.E.-type disease. Similar M.E. phenomena can occur due to diffuse CNS injuries from toxic chemical injury. I have seen this in police officers who have fallen into toxic chemical ponds in pursuit of those suspected of criminal activity. I have seen it in farmers repeatedly exposed to pesticides and herbicides, in hospital and industrial workers and in military personnel in contact with toxic chemicals, specifically toxic gases. I will discuss these at a later date as Secondary M.E. They do have one thing in common, and that is they also have a diffuse CNS injury as noted on brain SPECT scans. The diagnosis is made by history, as the actual cases are very difficult to diagnose due to the inability to assess brain levels of toxins in a live patient. Often these Secondary M.E. diseases are more severe than the infectious M.E. cases.

E. Caution: One should be careful in applying the diagnostic criteria discussed under the Nightingale M.E. Definition without also completing a thorough investigation. M.E., whether we are discussing primary or secondary forms, involves a significant diffuse injury of the Central Nervous System and an associated injury of the Immune System. This always implies the potential for secondary injuries or secondary diseases or pathologies caused by a dysfunctional brain and dysfunctional immune system. When the immune system is injured there is an impairment of the patient's ability to resist the development of malignancy as well as other important organ and systemic injuries.

F. Thyroid Cancer and Thyroid Atrophy: Due to funding limitations, we have demonstrated in our work only two characteristics of this corollary injury. The first is the high incidence of thyroid cancer in M.E. patients. In the general public, cancer of the thyroid occurs in 1-15 cases per 100,000. In our studies, in the case of the M.E. patient, thyroid cancer has an incidence of 6,000 cases per 100,000. For whatever reason, even if our figures represent some type of anomaly, the direction is obvious and suggestive of a major pathological association. We have already mentioned the pervasive vascular injuries. We believe that other pathological associations also occur. Failure to evaluate fully the M.E. patient may result in the physician missing important secondary pathology and possibly giving rise to patient death. **All M.E. patients as well as all chronic**

illness patients deserve a systematic and total body investigation. No individual should go through life, ill, disabled without knowing why he is ill. Simply offering a label, whether M.E. or CFS, without looking at the pathophysiology that gives rise to these disorders, is both unacceptable and potentially dangerous both for the patient and the patient's physician. (See *"The Complexities of Diagnosis"* by Byron Hyde, in the *Handbook of Chronic Fatigue Syndrome,* Eds. L. A. Jason, P. A. Fennell and R. R. Taylor. John Riley and Sons Inc. Hoboken N.J., 2003. This chapter is also available on various websites.)

G. Caution 2: Insurance companies regularly employ reputedly independent psychologists who demonstrate normal neuropsychological findings. Since the patient's data is unreliable if a test is done too frequently, the use of an insurance psychologist presents a grave problem in that neuropsychological testing by a truly independent Neuropsychologist may be delayed for up to a year before the patient can be properly tested. The conflicting results may tend to confuse any trial judge in a legal case.

H. Depression, Anti-depressive Medications and M.E.: M.E. is not depression; M.E. is not hysteria; M.E. is not a conversion disorder nor is it a somatization disorder; M.E. is an acute onset diffuse injury of the brain. Psychiatrists should not ever be placed in charge of diagnosis and treatment of M.E. patients. It is simply not their area of expertise and their meddling has at times caused great harm to M.E. patients. Also, during the 20 years that I have investigated M.E. patients I have yet to see a single case of real M.E. that has responded to psychiatric pharmacological treatment such that the patient has recovered and been able to return to work or school. This topic is a very large subject and demands a separate publication and this is not the place for it. However I would like to note again the vascular and cardiac pathologies that one encounters in M.E. patients and how M.E. patients are often made worse by one antidepressive medication that is considered benign. One of the most common antidepressive medications employed by psychiatrists and physicians in general for M.E. patients is an old pharmaceutical, Amitriptyline. Yet this medication may result in a condition referred to as *Torsade de Pointes*, a cardiac irregularity giving rise to resting tachycardia, QT interval prolongation and significant orthostatic hypotension. Since there is already a high frequency of these anomalies in M.E. patients, the use of Amitriptyline may assist sleep to some degree but may also simply worsen existing M.E. symptomology. I will hopefully return to this subject in another publication.

I. Graduated Exercise and the Myalgic Encephalomyelitis Patient:
Possibly due to the fact that some Fibromyalgia patients can be improved by a gradual increase in exercise, or possibly due to the so called protestant ethic that all you have to do to get better is to take up your bed and walk, some physicians have extended the concept of passive or forceful increased exercise to Myalgic Encephalomyelitis patients. This is a common and potentially dangerous, even disastrous misconception. Doctors Jay Goldstein and Ismael Mena, using Zenon SPECT brain scans, demonstrated that the physiological brain function of an M.E. patient rapidly deteriorates after exercise. They also demonstrated that this physiological dysfunction could persist for several days following any of several stressors. The physiological dysfunction occurs whether the activity (or stressor) is physical, intellectual, sensory or emotional. There are several problems with this finding. (1) The first is technological: Zenon is difficult to obtain and few nuclear medicine centres use Zenon. Nor is Zenon a dangerous substance, it is simply not used due to cost cutting. (2) Once the patient reaches a plateau, or starts to improve, lack of activity will eventually make the patient worse. Depending upon the degree of physiological brain dysfunction, patients should start to increase stressors slowly even if this means a temporary setback. This is neither an easy nor a fast process and again, depending upon the degree of brain dysfunction, may take years until the patient can resume a relatively normal life activity. (3) If the M.E. patient

conforms to the guidelines set out in this definition, the insurance company can only make the patient worse by instituting progressive aggressive forced physical and intellectual activity. M.E. is a variable but always, serious diffuse brain injury and permanent damage can be done to the M.E. patient by non-judicious pseudo-treatment.

J. Sleep Dysfunction: Many M.E. and CFS patients have multiple medical problems giving rise to their illnesses. Our office has in a few cases found up to 20 different pathologies and pathophysiologies in a single patient. The cumulative pathological weight is sufficient to cause any patient significant and chronic disability. One of many common problem areas is the nasopharynx and temporomandibular joints, a.k.a. the mandibular or jaw articulation. Several M.E. and CFS patients have significant pharyngeal and other obstructive airway problems that prevent adequate sleep function that in turn causes chronic fatigue syndromes and the associated chronic decrease in physical and cognitive stamina. Some of these correctable nasopharyngeal problems are so simple as to be mind-boggling. They include treating (1) enlarged tonsils that obstruct the respiratory tract when sleeping by surgery, (2) treating nasal obstructions, (3) treating chronic sinusitis with night time post nasal drip, and understanding (4) anatomically small pharyngeal box, (5) palate dysfunction and (6) temporomandibular dysfunctions that include mandibles that fall back to obstruct the pharynx when the patient sleeps. All M.E. and CFS patients should have a thorough investigation by an Ear-Nose-Throat specialist. Although it is costly, all M.E. and CFS patients should have a qualified orthodontist familiar with this group of illnesses carry out a careful examination of all M.E. patients. Unfortunately, sleep dysfunction testing and treatment is still at an early stage of its development. It is my experience that too often, when a sleep physiology physician finds a sleep dysfunction not related to obstructive disease or a movement disorder, he has little useful to offer in the way of treatment. Some sleep pathology physicians do go beyond this limitation and it is worthwhile for the treating physician to search for these rare individuals.

K. Viral, Hormonal and Age Related Triggers: I have discussed this briefly in the definition. This is a concept that is increasingly well known in medicine but to my knowledge has not been applied to M.E. Viral triggers are considered to be a possibility in certain asthmatic conditions, in multiple sclerosis, celiac disease and various rheumatoid conditions. All of these could be considered to be autoimmune illnesses. From examining hundreds of patients with fibromyalgia-like syndrome I cannot also wonder if **NSAIDS**, non-steroidal-anti-inflammatory drugs, that are increasingly prescribed for any pain condition, do not reset the CNS brain sensors to pain, thus creating chronic fibromyalgia and other pain syndromes. If this is proven to be true, then we can add pharmaceutical triggers that we already know can provoke rheumatoid disease. I have mentioned elsewhere the relationship of anti-depressive medication in causing or worsening heart dysfunctions, fatigue syndromes, sleep dysfunctions but they are not the only ones. Antilipid (cholesterol) pharmaceuticals appear to cause significant muscle weakness and joint and muscle pain in many M.E. patients, much more than in the general population.

L. Multiple Disabling Pathologies: Most M.E. patients have multiple disabling pathologies and it is insufficient for any physician who finds one pathology to assume he has found the one and only cause of this complex illness. Too often I have seen physicians who have found one major cause of M.E. or CFS dysfunction and illness, treat it, and then criticize the patient for not getting back to work when in reality, what the physician uncovered was only the tip of the iceberg.

M. Test Result Validity: When I was a medical student at the University of Toronto, our radiology professor insisted that as physicians, it was important to go over the actual X-Rays of the patient with the radiologist in order to develop an understanding of how to read an X-Ray and how to keep the radiologist aware of the pathology that you are investigating. Over the years I have had multiple reasons to visit a radiologist to assist me with reading routine X-Rays, complex intestinal X-rays, Ultrasounds, MRI and CT scans as well as brain

SPECT and brain PET scans. I cannot recall a single time that the radiologist did not take the time to go over the actual scans and X-rays with me and answer my sometimes very rudimentary and facile questions. However these trips to the hospital have also made me realize that the radiologist can miss major problems since they are not always aware of the individual patient's pathology. Recently, many SPECT scan and other technological facilities in Canada have simplified their technology, limited their findings to reports and failed to reproduce print-outs of their findings. This is true in Carotid and Transcranial Doppler examination where the velocity of blood flow through the arteries is not given, yet this is a valuable aid to understanding diseases related to arterial spasm. Yet the work sheets of the technicians contain this data. The same is true of the reading of EEGs. Neurologists too often simply say a test is normal since there is no evidence of a seizure disorder or a large space-occupying lesion. Often the neurologists go no deeper than this and miss major observable pathology. It is most unfortunate that so few centers have adopted **QEEG** or Beam technology, i.e. quantitative computer driven EEG technology. It gives significant better understanding of brain function abnormalities. The same is true of brain SPECT scans. These are very easy to learn how to read. I have already mentioned the problem with dropping Xenon scans. But recently, some Canadian centres have lost their experts in brain nuclear medicine and replaced them with individuals who are not expert in reading brain SPECTs.

They have also in some cases simplified the systems to maximize profit so that the detail is not always there. The hospital is paid the same for a badly done rushed SPECT as for an expert SPECT. This is increasingly a problem. For the physicians who only read the typed report stating, "the findings are normal" and who does not take the time to look at the brain images themselves, SPECT can be a useless exercise. I have mentioned the problem with thyroid ultrasound imaging. It is essential to insist that the radiologist actually give the measurements of each thyroid lobe rather than simply saying, "the findings are normal." This attention to detail is time consuming but also rewarding for the physician who is truly interested in understanding pathology.

Definition Changes & Improvements: As with all definitions, the Nightingale Research Foundation's Definition of M.E. will have to be looked at by many clinicians and researchers and increasingly knowledgeable patients and over the years, disagreed about, changed and improved upon. But what this definition does today is (a) separate clearly M.E. from CFS and (b) demonstrate that M.E. is an early diagnosable and provable disease - as are all true diseases, and (c) assist in the prevention and also the early treatment and cure of M.E. patients.

This Nightingale Research Foundation's Definition will be available with any updates or corrections, on the Nightingale Research Foundation's Website, http://www.nightingale.ca This definition may be copied, translated, distributed by electronic or hard copy and may be included, in whole or in part in any publication without permission from the Nightingale Research Foundation or the authors, provided that this last paragraph and referral back to our website are noted. A copy of any translation should be sent to Nightingale for possible inclusion in our website.

REFERENCES

Bastien, Sheila, chap. 51, pps. 453-460: Hyde, B., Goldstein, J., Levine, P. (Eds.): *The Clinical and Scientific Basis of Myalgic Encephalomyelitis / Chronic Fatigue Syndrome.* 1992, Ottawa, Nightingale Research Foundation Press. (**Note:** This publication is available on the Nightingale website.)

Diagnostic and Statistical Manual of Mental Disorders (DSM), 3rd edition (DSM-III), 1980, American Psychiatric Association (APA).

Fukuda, K., Straus, S.E., Hickie, I., Sharpe, M.C., Dobbins, J.G., Komaroff, A. L. *"The chronic fatigue syndrome: A comprehensive approach to its definition and study."* 1994, Annals of Internal Medicine, 121, 953-959.

Gilliam, A. G. *Epidemiological study of an epidemic, diagnosed as poliomyelitis, occurring among the personnel of the Los Angeles County General Hospital during the summer of 1934.* 1938, Public Health Bulletin, 240. Note: This publication will be available shortly on the Nightingale Website: http://www.nightingale.ca/

Holmes, G.P., Kaplan, J.E., Gantz, N.M., Komaroff, A.L., Schonberger, L.B., Straus, S.E., et al. *Chronic Fatigue Syndrome: A working case definition. 1988,* Annals of Internal Medicine, 108, 387-389.

Hughes, GRV. The antiphospholipid syndrome, A historical view. Lupus 1998; Supplement 2:S1-S4

See also: Sanna G., D'Cruz, Cuadrado M. J. Cerebral manifestations in the antiphospholipid (Hughes) Syndrome. Rheumatic Disease Clinics of North America, 2006, Aug;3.2 (3): 465-90.

Sanna G., Bertolacini, M.L., Hughes GR, Hughes Syndrome: A new chapter in neurology. Ann NY Acad Science, 2005 June; 1051: 465-86

Hyde, B., Goldstein, J., Levine, P. (Eds.): *The Clinical and Scientific Basis of Myalgic Encephalomyelitis / Chronic Fatigue Syndrome* 1992,

Nightingale Research Foundation, Press, Ottawa.

Hyde, B., "Cardiac and Cardiovascular Aspects of M.E./CFS," Chapter 42, Hyde, B., Goldstein, J., Levine, P. (Eds.): *The Clinical and Scientific Basis of Myalgic Encephalomyelitis / Chronic Fatigue Syndrome*. Nightingale Research Foundation, Press, Ottawa, 1992.

Jason, Leonard A., Fennell, Patricia A., Taylor, Renee R. *The Handbook of Chronic Fatigue Syndrome*, "The Complexities of Diagnosis," Chapter 3, Hyde B., John Wiley and Sons, Hoboken, New Jersey, 2003.

Montague, T.J., Marrie, T., Klassen, G. Bewick, D., Horacek, B.M., "*Cardiac Function at Rest and with Exercise in the Chronic Fatigue Syndrome,*" April 1989, Chest, Vol 95, p779-784,.

Magalini, S. I., Magalini S. C., *Dictionary of Medical Syndromes*, pps. 251-252.1997, Lippincott-Raven Publishers Philadelphia, 4th Edition.

Ramsay, A. M., *Myalgic Encephalomyelitis and Postviral Fatigue States* (2nd ed.) 1988, London: Gower Medical.

Rumack, Carol M., Wilson, Stephanie R., Charboneau, J. William, Johnson, Jo-ann M., *Diagnostic Ultrasound,* Third Edition, 2005, Elsevier Mosby.

Sharpe, M.C., Archard, L.C., Banatvala, J.E., Borysiewicz, L.K., Claire, A.W., David, A. et al., "*A report – chronic fatigue syndrome: Guidelines for research,*" 1991, Journal of the Royal Society of Medicine, 84, 118-121.

Sigurdsson, B., Sigurjonsson, J., Sigurdsson, J., "*Disease epidemic in Iceland simulating poliomyelitis.*" 1950, American Journal of

Hygiene, Vol. 52, 222-238.

Streeten, David H. P., *Orthostatic Disorders of the Circulation, Mechanisms, Manifestations, and Treatment,* 1987, Plenum Medical Book Company, New York and London.

Mental health problems in M.E. and Fibromyalgia Syndrome

INTRODUCTION

Since 1934 at least 70 myalgic encephalomyelitis (M.E.)-type epidemics have occurred around the world[1]. Yet most physicians and the public remain unclear as to the cause and characteristics of M.E. Many physicians debate the existence of M.E. as a valid medical entity. To put it kindly, many physicians simply find M.E. and patients with M.E. an unwanted bother. Since 1934, when the first well-documented M.E. epidemic ravaged the Los Angeles County Hospital[2], M.E. as a diagnosis has refused to go away and the mental health aspects of M.E. and chronic fatigue syndrome (CFS) not only remain but also appear to increase. We require a better understanding of M.E. and CFS. Even with knowledge, treating the mental health problems of the patient with M.E. will remain a formidable task.

Since 1984, I have been asking the question, 'What is M.E.?' First I sought out the few original M.E. experts in various countries who had examined patients with M.E. and the M.E. epidemics since 1934. I examined patients and questioned physicians from the 1934 Los Angeles epidemic, the 1947-48 Iceland epidemic[3], and the various UK, New Zealand, Australian and Canadian epidemics, and I visited each of these epidemic sites. For the next 24 years, I have intensively investigated patients with M.E. and asked: 'What are the pathologies of the patient with M.E. that cause them to remain ill and unable to carry out the tasks that they were so good at before their illness onset?'

During the past 24 years, I have confined my practice to patients with M.E. and CFS. As much as possible, I have examined every organ and system of each of thousands of patients with M.E., thanks to the totally free access that patients and physicians have to all tests and specialists in the Canadian health system[4]. The following chapter represents a small amount of what I have discovered during the years that I have questioned M.E. experts, studied the considerable M.E. literature, and examined thousands of patients with M.E. and CFS. What became obvious to me is that we cannot understand the patient with M.E. without examining the patient in great detail, along with his or her environment, social system and belief structure. We must also examine the limitations, usually imposed by government bureaucracy, that prevent physicians from adequately examining patients with M.E.

THE WRITER'S PREJUDICE

Allow me to start by first acquainting you with my prejudice of what I believe constitutes a patient with M.E. , since it may interfere with any preconceived perceptions that you bring to this discussion. My prejudice is this: the patient with M.E. has been with us a very long time, undoubtedly many centuries, but it was only in the twentieth century that we had the technology and medical organization to distinguish M.E. as a specific illness category and a communication system (the Internet) that enabled these patients to find each other and group together, often much to the regret of the physicians and often with an inaccurate diagnosis not based upon scientific evidence. A succinct definition of M.E. is this:
M.E. is an acute-onset, diffuse injury of the central nervous system (CNS) that in turn either provokes or is associated with organ, system and social pathologies that prevent the patient from effectively competing in their previous work and social culture.
Classically, M.E. occurs in epidemic and endemic periods, as in the 1934 Los Angeles, the 1947-48 Akureyri or the various UK epidemics. The onset of illness in both the historical and today's patients is often associated with an apparent infectious disease, immunization, and traumatic or toxic exposure in the immediate previous days or by repeated traumas in the prior weeks and months.

The key terms in this definition are:
- acute-onset;
- diffuse CNS injury;
- complex organ, system, and social pathologies.

- Figure 51.1 shows a single-photon-emission computed tomography (SPECT) brain scan of a typical patient with M.E. In the clinical situation, both the physician and the patient are confronted with another problem. Which patient has M.E. ? Which patient is misdiagnosed as having M.E. ?
- Possibilities include the following:
- Chronic patients with M.E. and its associated organ, system and social pathologies
- Chronic patients with undiagnosed or missed single or cumulative major medical illness or pathology diagnosed as M.E. but suggesting low-grade or slowly progressive injury
- Patients with classical psychiatric disease that may or may not be complicated by organ or system pathology
- Acutely ill patients misdiagnosed as M.E. but with a potentially treatable progressive illness.

The first three patient categories above tend to have similar mental and social health issues and can be discussed as a group. By definition, the referring physician never diagnoses any of the missed major pathologies. If they did, then probably the patient would not be referred to as having M.E. or CFS. The big danger, the veritable mine in the minefield, is the last – the patient who, either diagnosed by a physician or self-diagnosed, has recent-onset M.E. Too often have I seen fellow physicians who tell me they have M.E. , laugh and then say they do not require an examination, only to find that had they been examined properly an undiagnosed malignancy would have been discovered when it was still treatable. The greatest tragedy is to miss a diagnosis that could have been treated, and perhaps cured, by the physician who had taken the patient's symptoms not as a diagnosis but as a medical mystery to be solved by scientific testing. This also represents my major criticism of the diagnosis of M.E. , either by the physician or by the patient. My criticism is not directed toward the psychiatrist but to the physicians who, for whatever reason, have failed to properly investigate and follow these reputed patients with M.E. before referring them to psychiatry.

THE PATIENT WITH M.E. AND HER MYTHOLOGIES

For the clinical psychiatrist and clinical medicine physician, it is the patient's complex mythologies, perhaps more than scientific understanding, that overwhelm both the patient and the physician. It is essential to understand these patient mythologies in order to help the patient with M.E. Our patient is more than an injured body and brain: she is an integrated part of a complex belief and social structure with all of its values and prejudices.

If you were to be referred four real patients with M.E. , on average three would be girls or women and one would be a boy or a man. This 75 per cent percentage of females corresponds approximately to many autoimmune illnesses, such as multiple sclerosis and rheumatoid disease, but this of course does not validate M.E. as an autoimmune disease.

The child and adult sex distribution charts shown in Figures 51.2 and 51.3 were developed from an epidemiological study of approximately 2000 patients with M.E. or CFS investigated by the Nightingale Research Foundation during the epidemic period 1984-92. Note the significant divergence of females from males at puberty and the rapid drop in adult women new cases at menopause.

The patient before falling ill

From a mental health aspect, our patient with M.E. has a health history in addition to her present illness. Perhaps even more importantly, she has a vision of the future. The patient's past truly is prologue and the future is an existential necessity, both of which are capable of being destroyed. If you were to describe this young woman before the time when she fell ill, you would have acknowledged that she was a hard-working woman with many prior achievements and who contemplates realistic future goals. If she is in a long-term relationship, then in most cases she would have aspirations for her children, real or imagined, to do even better than herself. If our patient is a student, in most cases she will have all of the uncertainties, fears of youth, physical vigour and boundless energy, but also wonderful potential aspirations. Like Goethe's Faust, she has yet to learn that her existence is defined not by the goals but the very striving necessary to reach these goals.

Our patient before becoming ill will have already achieved a lot, including a higher education. Most often she will be a teacher or healthcare worker with one or more degrees. You will note a strong school and healthcare bias to this illness, suggesting an increased exposure to infectious diseases and with long hours of exhausting work. To a lesser degree, the patient's occupation will mirror the local employment population bias, but the school and healthcare bias will be paramount.

The occupations of 2000 consecutive patients with M.E. were tabulated during the epidemic period of 1984–92 (Figure 51.4). These patients came from across Canada, the USA and, to a lesser extent, the UK. Consequently, the figures are not prejudiced by local employment. By percentage, the single largest occupation was among respiratory technologies, followed by healthcare workers, including physicians, nurses and technicians in residential institutions for disabled people, which may have an increased infectious rate.

Often our patient will have been active in sports, with repeat associated minor and moderate physical traumas, often to the head and neck region. If you check her school record from kindergarten to the present, she will have missed a negligible amount of school. In most cases, before falling ill she will have had a history of excellent health, with no psychiatric associations. In my experience, except for those patients with a significant psychiatric family history, few of these patients with M.E. will be psychiatric patients. If you are an old physician like me, then our patient before falling ill will have been a person you would be proud to call your daughter. She will have an identity, courage, achievements and a belief structure with which you can associate. However, it is this belief structure as much as her illness that will provoke a significant part of what you perceive to be her illness. Thus, understanding her belief structure and her identity is important.

The patient's pre-illness identity and health belief structure

Our patient simply does not think about illness. If she is a parent, perhaps she has the natural concerns of any mother for her children, obtaining the right immunizations and paying occasional visits to the general practitioner (GP). She has never been significantly ill herself. Why think about illness? Visits to the doctor are for her pregnancies, routine Pap smear tests and breast examinations.

If you ask her about falling ill, she will dismiss the subject; illness is something short-term, a cold, perhaps off for a day or two and then back to school or work again. As a worker, she enjoys her job, the camaraderie, the striving; she loves her paycheque, even if it is too small – and it always is too small. She could be your daughter, your best friend's daughter or your granddaughter.

Life is simple for her. Should she fall ill and not recover sufficiently to return to work within a few days, she knows her GP will be able to diagnose and solve the problem with a pill or appropriate advice. Failing that, her GP will refer her to a colleague who will cure her and get her back to work.

If this first attempt fails, she knows that there will be a consultant or specialist out there who can appropriately diagnose, treat and cure her, but she never really gets to that concept, since the thought that her GP cannot treat and cure her simply never comes into her head.

Our patient has been working professionally or semi-professionally for over 15 years without ever falling ill. Let us assume our patient is North American. She believes she has taken adequate steps to ensure access to good and prompt healthcare. She has both short- and long-term insurance coverage deducted from her paycheque. She will have a mortgage, but she may not have taken out mortgage insurance, skimping on this to pay for her child's piano lessons. In any case, if she does fall ill for a few weeks or months, say from a compound fracture in a skiing accident in the French Alps, then her disability insurance will cover her expenses. No, she has never looked at her insurance policy. Why should she?

In the more than 2000 or 3000 patients with M.E. and CFS who have consulted me during the past 25 years in Canada, the UK and the USA, before their illness the majority had a long history of health, free of any serious medical or psychiatric illness; very few ever imagined themselves becoming ill with a chronic illness. Illness triggers are not always reliable as a cause of illness in all cases, but the data in Figure 51.5, taken from a large survey of almost 2000 patients, are perhaps reliable as rough casual indicators.

The disintegration of the patient's belief and identity structures

Our patient's carefully constructed mythology of enduring health until old age, and her perceptions of the infallible medical world, are about to be destroyed.

What will now confront our previous worker bee and future patient amounts not only to a total destruction of her most important belief structures but also to the ultimate loss of her identity, an identity carefully constructed since youth.

The young woman falls ill. Her illness usually has an acute onset, but it is not short-term. From then on, nothing works, nothing unfolds in the manner she might have perceived it before the onset of this strange illness.

She does not become better in 2-3days. In almost all cases, her doctor does not appreciate that he or she is dealing with the onset of a chronic illness. Since most acute illnesses tend to resolve on their own, her GP tells her that she will be fine in another week. That does not happen: her illness persists. Then or at the next appointment, her GP may prescribe a medication or treatment and may send her for a few tests, usually a complete blood count and a urinalysis and not much else. The tests tell her GP nothing because he or she does not know what tests to order; nor does the GP know at this point what he or she is dealing with. Sooner or later her GP will refer her to a consultant.

'No findings of note,' the reply comes back. 'Perhaps your patient is depressed?'

If her GP has not made the diagnosis by now, he or she soon will. It is likely to be anxiety neurosis, depression, work avoidance or family dispute – all diagnoses that are safe to make, all of which have no proof. Often notations in the history are not told to the patient; these notations, when sent on to the next doctor or other agencies, may undermine the patient's later chance for insurance benefits or state incapacity or disability allowances. The doctor is thus protected, but not necessarily the patient. Our patient may even be given an antidepressant that, over time, succeeds only in significantly increasing her weight. Usually the patient is given a psychiatric diagnosis by the GP and referred to a psychiatrist, although there has been no previous psychiatric history. At times she may believe her GP that she is depressed; more often she does not. Even if the doctor diagnoses M.E. , he or she is unable to treat M.E. adequately. There simply is no magic pill and no remedy, and there is very little valid knowledge about the significance of the diagnosis. The psychiatrist immediately is at a disadvantage: he or she believes that physical disease has been ruled out. Yet physical disease has not been ruled out, and in most cases it has not even been seriously investigated. Our patient is now labeled 'M.E. or CFS', but often the doctor thinks anxiety neurosis, depression or conversion disorder. If the doctor thinks CFS, then this is no more than a poorly defined label; if he or she thinks M.E. , then this may be the least properly investigated epidemic and endemic illness in existence. In the UK, very few of these patients so diagnosed have ever been subjected to any significant scientific investigation; many have not even had a serious physical examination.

Weeks and even months go by. No medication and no treatment works. Nothing works and the illness persists.

Within 2 months, her disability insurance raises its head. Since there is no documented evidence for any significant illness, her insurer does not bother paying. Sweet soul that she is, she honestly believes that if the insurance company only knew how ill she were, then it would send her a disability cheque. The insurance company has a good cop who pretends to be the patient's friend, talks her into going back to work or tells her the company is just waiting for a few more results to be sent in. She tries to go back to work and falls flat on her face. Before long, a year has gone by and in many cases the insurance company still has not honoured the disability pension. Some insurance companies actually guide the patient along until it is outside the time limits for her to make a qualified application for benefits. By now, much if not all of her savings have been dissipated. All ballast, including her child's music lessons and the piano, have gone. Rapidly she is stripped to the economic bone. If she has not been frantic up to now, she soon will be.

The search

Our patient, worker bee that she is, after months and possibly a year or two, having failed to find a helpful doctor and failed to have been awarded her disability pension, searches the Internet for an expert who will treat her. Her GP and often her family and friends have given up on her. Yet she cannot believe that there is not someone out there who can diagnose and treat her and get her back to work, give her back her family, paycheque and friends, and restore her identity. She amasses reams of garbled information, which is often misinformation – but she does not believe this. She is going to educate her doctor about the 'true facts' about M.E. However, on the Internet, she also encounters mixed in with the fact some of the most incredible hogwash imaginable; she will meet some of the slickest charlatans in the business, the penny-and-pound thieves who sell her miraculous alternative medications that do not work and sometimes kill. Before long, our disabled patient is paying out, in effect gambling, incredible sums of her much-needed money in the vain hope that these miracle cures will soon make her better and get her back to work. They do not work.

Some of the more brilliant charlatans, the highwaymen of the twenty-first century, will guarantee to restore her health: 'Our treatment has saved thousands of people, but this treatment is expensive, very expensive, but isn't your health worth it?' She bites! If she or her family are desperate enough, they will sell their house, invest their life's income in a desperate gamble not to lose all. They lose all! Sometimes she will lose her family; but worse, she will lose her carefully crafted identity as a proud member of her community, a successful worker and a student of life. She loses all that she was and intended to be. She cannot even be a good mother. She thinks about the piano lessons that have stopped and the piano that has moved out, and she feels that she is now failing as a parent. Nothing works.

Several times she tries to go back to work but simply cannot endure. Since she has no adequate psychiatric or physical diagnosis, her long-term insurance policy has either stopped after a short while or in many cases has never begun. The insurance companies know these patients well. They know their client has no funds, no physical energy and no resources to fight back. Our patient's partner may have left her; he can no longer take the expense or handle chronic illness, the lack of sex, his wife's inability to even go out for a show. Alone, or with her children, she may return to live with her ageing parents. Without supportive parents, she may fall into a life of welfare benefits and despair, barely eking out an existence. She may kill herself. Only if she is one of the wealthy few, one with funds that are not wasted, or who succeeds in obtaining her disability pension is our patient able to survive with any integrity. Many patients become bitter, writing diatribes, attacking authorities, doctors included. Many doctors simply lump them all into one more group of nutters.

She will have been referred to you, a psychiatrist. You are kind to her. You speak in a soothing manner. Having lost everything, at last she has found a friendly ear, someone who will believe her. She cries – a sure sign of depression – and out comes the prescription pad for the first of an endless series of antidepressive medications that succeed only in making this normal-looking athletic woman into a fat blimp and still unable to return to work. If our patient resists this treatment, she is written off as 'not complying'.

DOCTORS: THEIR FEAR OF EMBARRASSMENT AND THE ECONOMIC REALITY

Let me introduce you to a few of the doctors that this woman has consulted. Like our patient, these doctors have their own history and mythologies.

Most doctors were some of the top students in their primary and secondary school systems. Then they were enrolled in medicine, where they met hundreds of other medical students with a history of being at the top of their class. Only a few of the hundreds of these medical students can ever expect to be at the top. Only a handful would be considered to be the most brilliant and have honours bestowed upon them. No one doubts that the top students would be truly brilliant. But, depending upon the school and the teacher, their brilliance may in large part be the ability to give back a wide range of accepted fact in a clear, concise, organized manner. It is a curious thing that many of the very top students do not continue in clinical medicine. They are not always interested in people as much as they are in excellence. The majority of the other graduating doctors in their class fall into other groups, and those who have not lost their curiosity may become some of the very best clinical physicians and researchers. Yet others who have always been first, the best in their class, have been needlessly embarrassed at not being one of the exalted top few. For many such students of medicine, this may be the end of independent thinking, of exploring, of challenging the accepted wisdom. It is precisely this need not to be embarrassed further that caused doctors for centuries not to discover circulation or the concept of infectious disease and almost every modern aspect of medicine. For centuries, they survived by embracing the accepted wisdom.

There is another problem concerning the referring GPs who will refer the patient with M.E. to you. During the past 40 years, except for a few subspecialties, most doctors' real taxable incomes in North America, the UK and Western Europe have fallen from one of the highest in the community to that of a modest middle-income person. You may be earning less than your patient with M.E. was earning before she became too ill to work. The doctor starts in practice with debts. Without help from their parents, in most cases there is no way that doctors in North America can afford to buy the house that any doctor purchased 40 years ago or even 20 years ago. The medical magazines talk about making your practice more efficient, which essentially means seeing more patients for less and less time. Essentially, that means getting the patient out of the clinic by giving them a pill, sometimes any pill. They and you pretend this is not so, but you too have a mortgage and your partner expects you to buy that magical piano, pay for your child's piano lessons and the family trips to ski in France this winter, or simply take that cruise advertised in the doctors' magazines. Unless you take a government, pharmaceutical industry or insurance company job as a doctor, or you succeed in getting one of the senior positions in the hospital on a salary, unless you work a 60-hr. week or see patients outside the National Health Service (NHS), not only will your income be modest but also you will be judging yourself against an unrealistic measuring stick of those doctors who have come before you and a few of your colleagues who appear to be wealthy. You simply do not need a troublesome, complicated patient with M.E. who will take up your valuable time.

I am still talking here primarily about the non-psychiatrist physician. Money is status. Time is money, and few can take the time to explore the incredible significance of the relatively young chronic patient with diffuse brain injury. These patients with M.E. simply take too much time. Consequence? The primary care doctor simply sends the patient with M.E. to the psychiatrist, who they believe has all the time in the world. The psychiatrist in turn will assume that, as in all reputed psychiatric patients, the primary care doctor and the consultant have adequately investigated the referred patient who now sits in front of them. Yet you, as a doctor, should know better – and many do. The system in the UK prevents most primary care doctors from even ordering a full technological evaluation of their patient. The system in the UK, North America and much of Europe does not give the doctor time or money to explore this fascinating patient. Worse, there are no specialty clinics working inside the NHS that have the time, authority or financing to properly investigate these patients with M.E. for physical cause of disease. This total chronic failure to systematically examine the patient with M.E. has not been helpful, either to the patient with M.E. or to the psychiatrist.

The relatively young, chronically physically ill public in the UK tend to be very critical of the psychiatrist, and often needlessly so. Yet, as a psychiatrist, you have both a disadvantage and an advantage over the regular non-psychiatrist. The disadvantage is that your income may be one of the lowest in the medical community. The advantage is that your expenses are less and you alone among your colleagues will be able to listen to your patient for a considerable time. Perhaps you will be the first doctor who has the time to listen to her and who can get to the physical route of her anxiety, even if you cannot improve her material life. It is here where your difficulty becomes even more complicated.

Like many of your medical and psychiatrist colleagues, you will not want this patient. She simply takes up too much time in the clinic, and you wish to see her depart so that you may get on to a treatable patient. Many of your own psychiatrist colleagues will believe that any patient with M.E. is simply expressing some form of hysterical behaviour, is a whiner, is someone who does not want to work, or, worse, is boringly uninteresting. The patient will believe she knows much more about M.E. than you do; perhaps she does, but that is irritating to you and often her information is as highly inaccurate as yours has been. We are now looking at a major treatment impasse, including hostility, which may be mutual, and poorly organized information on both sides.

There is one final aspect to these patients with M.E. Few patients realize the two-way nature of medicine, the great joy of being a doctor, of making a clear diagnosis of the patient and getting that ill patient either back to health or at least to a position where they can manage. Yet this patient with M.E. is not some fascinating patient with bipolar disorder from the Bank of England who has just ripped through a few billion pounds of your country's money. She is not a schizophrenic poet of immense talent and immense self-destructive powers. This is not a case where there is a good chance you can bring banker or poet back to reality. It is so easy to tell yourself that this is one more case of M.E. hysteria.

How can you, the psychiatrist or for that matter any doctor, help this young woman? If you catch the patient in time, you can attempt to help this unbelieving patient to stop wasting her valuable and decreasing funds on bogus care and treatments, usually from non-doctors. If you believe that the patient is significantly disabled, and in my experience few patients in this category ever lie and most are more disabled than even they realize, then you can help her obtain appropriate state benefits or her entitlement to disability insurance. More than any other assistance you can provide, this may help her save part of her identity and perhaps her life, and she will bless you for it forever. This may take work, and it would be good to have a psychiatrist colleague or ombudsman whose work is limited to handling these matters to which you can refer this patient. It takes both significant time and skill in assisting these patients with the insurance industry and benefits systems.

It is not only the uninformed doctor who represents a problem but also the reputedly informed patient. Both doctors and patients have bought into the nineteenth-century Oslerian principle that the best way to treat and possibly cure a patient is to diagnose the illness as to the cause, treat the cause and, with knowledge and luck, cure the patient. The problem is the phrase 'the cause', since many of these patients have multiple causes giving rise to their disability. This unicausal theory of medical pathology worked for pulmonary tuberculosis, for syphilis, so why not M.E. ? Well, what if there is not a single cause of the illness? What if the patient with M.E. is disabled due to multiple cumulative pathologies?

MULTIPLE PATHOLOGY PATIENTS

My clinic is in the process of an in-depth study of the last 53 consecutive patients referred to us with a diagnosis of M.E. or CFS. In this group:

- 100 per cent had missed measurable brain dysfunction;
- 98 per cent had measurable significant sleep dysfunctions that included (i) lack of type 3 and 4 sleep, (ii) abnormal, absent or significantly delayed rapid eye movement (REM), (iii) central and peripheral apnoea, (iv) restless legs syndrome and (v) oxygen saturation that fell below 88 per cent (interestingly, oxygen saturation below 88% caused loss of consciousness in an aircraft pilot);
- 74 per cent had measurable thyroid dysfunction;
- 47 per cent had measurable significant arthritic, rheumatoid changes or other indicators that were previously diagnosed as fibromyalgia;
- 47 per cent had other missed major disease, including cardiac disease, malignant disease, vascular injuries and
- autonomic nervous system dysfunction;
- at least 16 per cent had typical psychiatric illness but, in addition, missed physical disease.
- The problem here is not the multiple pathologies but the fact that the primary or consulting doctors missed these multiple diagnoses.
- We subdivided the 53 patients in this study under occupations. This is what we found:
- Nineteen patients (37%) were school-associated professors, teachers or students.
- Nine patients (17%) were civil servants, the majority with young children in school.
- Five patients (10%) were healthcare workers.

This study of a group of 53 patients with M.E./CFS was biased due to the fact that in Ottawa, Canada, a large percentage of the inhabitants are government workers. In an earlier study that looked at 2000 patients from across Canada and the USA, school and healthcare workers represented over 70 per cent of the 2000 patients. This suggested that M.E. was usually associated with a high exposure to infections.

As noted in the group of 53 patients, the single largest group was 19 (37%) of the total. These were students and teaching staff at primary, secondary and university educational institutions. For brevity, I will confine this discussion regarding pathological findings to these 19 patients referred to me as patients with M.E. who were teachers and students. This dominant group also contained the largest number of youths and children.

Collectively, these 19 school-associated patients had been seen by over 200 doctors. All had missed the following pathologies, which were found by extensive history, physical and technological examinations:

Dysautanomia and postural orthostatic tachycardia

Check your patient's arterial blood pressure, pulse pressure and heart rate when they are lying, sitting and standing. Have your patient stand without moving and check their blood pressure every minute for 10-15 min, stopping if their pressure drops precipitously or their heart rate accelerates above 120 bpm. If you find the blood pressure falling or the pulse pressure narrowing significantly, immediately stop the test and ask the patient to sit down. Make one of your first patient visits a 2- to 3-h visit in order to obtain a full life history. Note how many times the patient has to go to the toilet during this visit, suggesting possible pituitary or other endocrine disease. Do not allow the patient to have a water bottle with her during this time.

Measurable brain disease in patients with M.E.

A surprising 100 per cent of the 19 teachers and students had significant brain changes or anomalies by one or all of technical examination, measurement or history. If you can, order or have your medical colleagues order a brain SPECT, a magnetic resonance imaging (MRI) scan with contrast of the brain that includes pituitary, cerebellar tonsils and cervical spine area.

As mentioned earlier, the following pathologies were missed by over 200 examining doctors, who possibly did not believe they were dealing with significant physical disease and so did not do an extensive examination of these patients. We found the following:

- A patient with tertiary CNS syphilis and who was also positive for hepatitis B diagnosed as having simple major depression
- Leucoencephalopathy with ventricular hypertrophy in a youth
- One adult with significantly abnormal electroencephalograms (EEGs)
- One youth with missed nocturnal seizures and seizure-associated episodic complete heart block, causing syncope, significant hypoglycaemia and obstructive chronic tonsillitis that occluded his pharynx when sleeping on his back
- A youth with Chiari syndrome with ventriculomegaly
- An older patient with significant generalized brain atrophy and ventricular hypertrophy
- A patient who had been in the area of the Chernobyl disaster as a 1-year-old child and who had a subarachnoid cyst that had displaced two-thirds of the left hemisphere, including the entire left frontal lobe, the entire left temporal lobe and the anterior part of the left parietal lobe, but who had no observed neurological examination abnormalities either on neuromuscular examination or in gross intelligence. He simply had overwhelming exhaustion. He had graduated with a master's degree
- A patient with complete atresia of the middle cerebral artery
- A patient with multiple CNS vascular changes.

In addition to these findings, many patients had major SPECT brain changes in both hemispheres, the midbrain and the brainstem.
The average age of these 19 students and teachers was 33 years.
Two of the 19 patients had incapacitating autonomic nervous system dysfunction. One, a master's student who fell ill immediately following recombinant hepatitis B immunization and is now house-confined, has a highly positive tilt table test and is unable to maintain her blood pressure at a physiologically normal level while standing or on movement.

The rewards in properly investigating this group of patients are significant. Although we did not find any patients with multiple sclerosis (MS) in this group of 19, we did pick up a missed case of MS in the total group of 53 consecutive patients and a surprising number of single large (diameter ≈ 2 cm) CNS demyelinating lesions that do not qualify as diagnostic of MS. Each of the patients with non-MS single-lesion demyelination was associated with markedly abnormal SPECT brain scans. Also in the group of 53 consecutive patients, but not in the school-associated group, we found a missed significant brain aneurysm in addition to numerous other organ pathologies in the same patient. The aneurysm has since been repaired. Another patient had a right lenticular haemorrhage, which the neurologist then dismissed as minor; on SPECT we were able to demonstrate a 3- to 4-cm halo of abnormal activity around the lenticular lesion consisting of highly significantly hypoperfused brain tissue. In addition, we observed a decrease in perfusion in the entire right lobe of this patient, but with a normal left hemisphere perfusion. Clearly some localized non-motor cerebral accidents may provoke profound generalized CNS changes.

Are these all patients with M.E. ? No, of course not; but in real terms, it does not matter. These individuals were all diagnosed as having M.E. or CFS by otherwise competent doctors who dismissed the patient when they came with a diagnosis of M.E. These will be the same people who are referred to you as patients with M.E. by GPs and internists who simply think these patients complaining of acute or gradual-onset fatigue and cognitive dysfunction have M.E. or CFS and either have no idea how to investigate them or simply are unwilling to take the time to do so.

Thyroid disease in patients with M.E.

For some reason, the centuries-old medical knowledge of the physiological association of thyroid disease and intellectual, emotional, psychiatric, cardiac and other endocrine pathology seems to have escaped the 7min-per-patient visits of many primary care and specialist doctors. Most non-psychiatrist doctors limit their examination of the thyroid to a cursory palpation of the gland for nodules and perhaps order thyroid-stimulating hormone (TSH) and free thyroxine (T4) tests. In most cases these doctors will not find disease. It is essential to request a thyroid ultrasound on all patients with M.E. In addition, the doctor must ask the ultrasound technician to give the measurement of each thyroid lobe; if you do not, the technician or the doctor reading the ultrasound result will often just say 'normal' if there are no nodules or the gland is homogeneous. Why the measurements? Simple: the Mayo Clinic normal thyroid sizes are 13-21 cm^3 for females and 15-23 cm^3 for males. They arrive at these figures by multiplying together the three dimensions of each lobe and adding the left and right lobe measurements as though the thyroid were a regular rectangle. This is not the actual size of the thyroid. If you have access only to an old ultrasound device, then the radiologist may give you these crude rectangular volumes. It is necessary to multiply these volumes by a factor of 0.51. This will give you the approximate normal thyroid volumes of two irregular solid spheres. The real volumes are then half the Mayo suggested volumes, or 6.0-10.5cm^3 for females and 7.5-11.5 cm^3 for males. It is important to know this, since modern ultrasound machines give this calculation in terms of the 6.5-10.5cm^3 female thyroid volume scales. These are not gold standards, but if you bring in a thyroid at less than 4 cm^3 or over 15 cm^3, you know you are dealing with an atrophic or hypertrophic thyroid, respectively, with possible intellectual, emotional and psychiatric consequences. Sometimes, simply by appropriately treating these patients with T4 or tri-iodothyronine (T3), you can cure their fatigue and cognitive dysfunctions. (Note: patients on previously prescribed or over-the-counter thyroid medications may have a hypotrophic thyroid.) Nor can you count simply on the usual TSH, free T4 (FT4) and free T3 (FT3) thyroid tests that are stated as normal or slightly abnormal. If the basic injury to the thyroid is vascular, as I believe it to be in most M.E. brains, then often the parathyroid hormone (PTH) and the ionized calcium will become abnormal before the usual thyroid tests. I do all of these tests, including thyroglobulin, thyroglobulin antibodies, and microsomal antibodies, as well as an ultrasound, on each and every patient with M.E. that I see. In the group of 19 teachers and students in our group of 53 referred patients with M.E. , we found 14 patients (74%) with abnormal thyroid activity. I have also found several patients with new-onset M.E. with normal initial thyroid chemistry but with a significantly shrinking thyroid when the ultrasound was repeated in 1-2years, which may suggest a vascular problem.

In the last 100 patients with M.E. /CFS, we also have found missed thyroid malignancies in 6 per cent. Some doctors discount this figure, since thyroid malignancies are considered to be common and not particularly dangerous, but that is not true for a young population. In one of the group of 53 patients, we discovered a thyroid malignancy that had already disseminated.

There is one more thing you must know before we leave the thyroid. T4 does very little on its own: T4 must first be discharged into the bloodstream and then transported to the liver and kidneys, where one of the iodines is removed to make T3. If there is pathology in this conversion, then an isomer called reverse T3 is made. In simple terms, the importance of this is paramount. T3 is one of the keys that turn on each body cell. If your patient is producing too much reverse T3, then this reverse T3 fits into the cell's energy receptor and breaks off, and the cell energy cycle cannot function normally. All the T4 in the world will not help this patient: she requires T3. Is it enough to ask for a reverse T3 level? No! Often the laboratory will give you a reverse T3 level in the normal or high normal range. You need also to order an FT3 at the same time and then divide the reverse T3 level by the normal T3 level; if the result reaches 10 per cent or more, your patient probably requires exogenous T3. But be careful: these patients with M.E. tend to be very medication-sensitive, so you should start at 5 mg daily, increasing every 2-4 weeks until you reach 25 mg and then stay at that dosage for some months before considering raising it to a normal dosage or subnormal dosage. If the patient has a heart condition as well, take advice from a cardiologist on the safety of giving T3 to this patient. All of these patients want to get back to normal in 5 min, but this is dangerous if part of the problem is thyroid dysfunction. Starting T3 at too high a dosage or increasing the dosage too rapidly may provoke seriously irregular heart rates, irregular cardiac rhythm and rapidly altering blood pressures. It may take up to a year or longer to slowly restore the patient with M.E. to normal thyroid levels.

Sleep dysfunction

All patients with M.E. should have at least one sleep study and some sleep studies with film monitoring to observe the presence of nocturnal seizures missed in daytime EEGs. In our group of 53 patients, we found only one patient with a normal sleep study; her dysautanomia was so severe that she was in a state of vascular collapse. Some authorities state that the abnormal sleep study can be blamed on the way the study is done, or the hospital environment in which these tests are performed, but generally these patients have non-restorative sleep. In our studies, 70 per cent of the 53 patients had no stage 3 or stage 4 sleep. This is the sleep phase where short-term memory is laid down. In addition, 77 per cent had grossly insufficient REM and very long REM latency. Among other functions that occur during REM is the burning into the neuron system of short-term memory: it is no wonder, then, that these patients describe short-term memory loss.

Interestingly, if the oxygen saturation falls below 92 per cent, commercial pilots become both colour- and night-blind and have difficulty landing their planes during the night. If the oxygen saturation in the cockpit falls below 88 per cent, the pilot has a good chance of losing consciousness and crashing the plane. Accordingly, oxygen saturation is monitored carefully in the cockpit, although less so in the passenger compartment. In our study of 53 patients, the oxygen saturation in 25 patients (53%) fell below 92 per cent during the sleep study; the oxygen saturation in 17 patients (30%) fell to 88 per cent or below. In other words, they were not sleeping: they were unconscious. The oxygen levels at least are potentially correctable pathologies.

Respiratory dysfunction

In our group of 19 students and teachers, ten patients (53%) had some measurable respiratory dysfunction. We did not count a history of asthma in this group, although in many patients there was an obvious overlap. Had we included a history of asthma, it is possible that even more patients would have had measurable respiratory dysfunction.

Missed miscellaneous disease

In our group, 47 per cent of patients had major missed illness, including the following:
- A case of tertiary syphilis and hepatitis B
- Four patients with significant heart disease
- A patient with respiratory dysfunction with significant pulmonary valve disease
- One juvenile and two type II cases of missed diabetes
- One patient with Ehlers-Danlos syndrome
- Numerous rheumatoid and significant spinal anomalies.

Although two patients had significant incapacitating autonomic nervous system dysfunction, several had lesser degrees of measurable dysfunction.

One patient in the group of 53 patients was a young lawyer and Olympic runner. He was referred to me as a patient with M.E. due to extreme fatigue and had Marfan-like anatomical changes. He had a missed hyperelastic distended thoracic aorta picked up on echocardiogram. In addition, he had lumbar dural ectasia on computed tomography (CT) scanning (seen in 65-92% of patients with Marfan's syndrome). At least seven doctors had each missed this diagnosis. The patient was denied any family history of Marfan's syndrome until he was asked to enquire of his remote cousins. Three of his second cousins had a family history of ascending aorta surgical replacement in their twenties. The cause of his thoracic aorta pathology was complicated by the fact that he had been infected with brucellosis while training in South Africa. Brucellosis can also cause aortic aneurysm. A detailed extended family history is essential in investigating M.E.-type patients.

Psychiatric disease

Patients with M.E. like to believe that there is no psychiatric disease associated with M.E. Some psychiatrists and primary care and specialist doctors like to believe that 100 per cent of patients with M.E. have psychiatric disease. What did we find? Many of these patients with M.E. had been referred to psychiatrists before our examination of them; six patients (32%) were diagnosed with primary psychiatric disease. After we had finished our investigation, we found only three patients (16%) with treatable psychiatric disease. There are several reasons for this discrepancy. We believe that some psychiatrists, to assist a destitute patient with M.E., may give a psychiatric diagnosis simply to assist the patient in obtaining their disability pension.

Let me give you a brief review of some of these psychiatric patients.

The patient with tertiary syphilis was first diagnosed with unipolar or major depressive disease. This patient also had hepatitis B. Another patient had major childhood abuse, during which time she was placed in a reformatory and also various psychiatric hospitals in Switzerland by the woman who had adopted her as an infant. This incarceration first occurred when the patient reached puberty and may have been due to the mother's sexual jealousy of having another female sharing her ambassador husband's innocent affections. On her own resources, my patient obtained her master's degree, became a secondary school teacher, and is fluent in English, French, German, Italian, Spanish and Latin. There was obvious trauma. Today, aged 53 years, she is exhausted, perhaps simply worn out, and she certainly has anxiety and depression.

A third teacher, also adopted, initially diagnosed as having depression, has serious food addiction, which in itself has caused multiple medical problems. He certainly has no overt psychiatric illness, but he does have brain dysfunction and memory disorder. He also had missed generalized vascular disease, missed diabetes and missed myocardial infarct and is one of the two patients in the school group with generalized atrophic brain syndrome.

The husband of another patient, also a teacher, committed suicide; the patient has made two attempts on her own life and was reasonably diagnosed as having unipolar or major depressive disease.
Two other patients were diagnosed as having unipolar depression. Both had seronegative rheumatoid arthritis and both had measurable significant heart disease.

Each of these six patients also had other major physical pathology. How many of the six were truly classical psychiatric diseases? Probably three, or four if you count the teacher with brain atrophy and cognitive dysfunction. What is most incredible is the truly remarkable resilience of some of these patients despite their multiple organ and system disease.

Do these 53 consecutive chronically ill patients have M.E. ? Is that important? Doctors who believed that the patients had M.E. or CFS referred all these patients to me. The same patients will be referred to you with the same bewildering array of pathologies. Can you afford to dismiss them as psychiatric patients if they have not been properly investigated first? Since all of the definitions of CFS and to a lesser extent those of M.E. are based upon symptoms common to a multitude of serious pathophysiological illnesses, under the present understanding of M.E. and CFS, doctors, whether primary care, consultant or psychiatrists, cannot simply dismiss these patients without first documenting an extensive investigation of the individual patient. A thorough investigation of the chronically ill younger patient is fundamental to good medicine.

DEFINITIONS OF M.E. AND CFS

You may remember the lines of Mathew Arnold's poem 'Dover Beach'. M.E. and CFS definitions are a bit like that.

And we are here as on a darkling plain

Swept with confused alarms of struggle and flight
Where ignorant armies clash by night.

    The reason that I did not start this chapter with a definition of M.E. is that there is no general accepted agreement on the definition of M.E. , or the pathophysiology of M.E. , or an accepted understanding of M.E. chronicity. Many doctors simply do not believe that M.E. exists except as one of various psychiatric or social disorders or as an Internet-constructed example of mass hysteria. One would think that this was bad enough, but it gets worse. There is disagreement as to whether M.E. and CFS represent the same disease spectrum. There are now thousands of scientific publications on M.E. and CFS, but those publications on physical dysfunction are much like the parable of the six blind wise men of Hindustan, who individually described the elephant as a wall, a spear, a snake, a tree, a rope and a fan. To my knowledge, since and including the 1932 M.E. epidemic in the LA County Hospital and the first excellent publication of that epidemic by AG Gilliam, no one has ever done a complete long-term systematic scientific study on any significant number or patients with M.E. or CFS. No wonder, then, that doctors and scientists are in disarray on M.E. and CFS. It is also obvious that many people diagnosed with M.E. and CFS are missed patients with multi-system, multi-organ pathology. Are these pathologies due to a specific CNS injury or generalized injuries acquired during the initial infectious, immunological, chemical or traumatic injuries, or are they genetically related illnesses? We simply do not know.

    Nor is there any agreement as to whether M.E.[5] represents the same disability as CFS[6,7]. In the USA, the conflicting name for CFS is 'chronic fatigue and immune dysfunction syndrome' (CFIDS). In the UK, some doctors have used the term 'postviral syndrome' to describe M.E. In addition to the above definitions, there are possibly better, more recent definitions, including the so-called 'Canadian definition'[8] and the 'children's definitions' by Jason and colleagues[9,10]. Unlike any other medical condition, real or imagined, I know of no other where there are two warring armies of very well-educated doctors who are so critical of the others' views on the subject as to whether this is a real or imagined illness or whether M.E. is the same as CFS. There should be no surprise when I tell you that there is no accepted agreement on the treatment of M.E. or CFS either. This treatment quandary is particularly true if you examine the complexity of the pathologies hidden within the group of patients I have discussed above and whom well-educated doctors refer to as having M.E. and CFS. Let us briefly discuss the definitions.

Definitions of reputed infectious and other diseases come in various sizes and constructions, but essentially there are two types of definitions:

• Epidemic-based definitions: these are from the bottom up and are based upon epidemic findings when large groups of individuals fall ill at the same time, usually in confined quarters. The clarity of such a definition changes with technological and clinical advancements, particularly if a single causative agent is found. Physicians and researchers who investigate the epidemics always construct these definitions. These definitions, often faulty, at least have the benefit of being based on physical findings and follow-up of actual patient illness by doctors and scientists who investigate the actual patients over a period of time.

• Theoretical-based definitions: these are based on a theory of preconceived limits of what an illness is supposed to be and what is supposed to happen in the illness. They are often based upon symptoms rather than pathological findings. The Centers for Disease Control and Prevention (CDC) definitions of CFS are typical of this type of bureaucratic definition. This type of definition excludes all patients who do not follow this hypothetically derived definition. This type of definition is often bureaucratic, dictated from the top down, and those that direct this definitional process at times tend to have little and often no experience in primary patient investigation of the illness process that they are attempting to describe. Unfortunately, this is not unusual in medicine today. Since the same symptoms are often common to a multitude of different illnesses, these definitions tend to be misleading.

Why is the derivation of a definition important? Before Pierre Marie and Ivan Wickman's description of epidemic poliomyelitis based upon the first major poliomyelitis epidemics in 1887 and 1895 in Stockholm,[11,12] what we now know as poliomyelitis was described under a multitude of different diseases, with multiple different names and different causes. Epidemic descriptions by clinical investigators with careful follow-up have the advantage of bringing together the multiple and varied aspects of a single illness. One eminent French neurologist in the 1890s described the Scandinavian polio epidemics discussed by Wickman as examples of mass hysteria, although he had never been to Scandinavia or examined any patient. It was only when doctors such as Wickman studied the outcome of the late nineteenth-century Scandinavian poliomyelitis epidemics that doctors and scientists were able to bring these previously multiple disease phenomenon together under one classification. Unfortunately, in none of the more than 60 epidemics of M.E. has anyone thought to do a funded systematic pathophysiological investigation and long-term follow-up of these epidemic patients. This itself is perhaps the biggest tragedy.

Definitions of M.E.

Doctors and workers who studied patients during epidemics of M.E. developed the following definitions. The definitions were not good, since most were developed at a time when virology was in its infancy or later when scientific research was largely underfunded:

- Onset and location: these were acute, epidemic and concurrent endemic episodes that occurred in both children and adults starting in late summer and early autumn in the north temperate zone. Onset of new disease tended to decrease rapidly after October, finally trailing off during the Christmas–New Year period. Most frequently described epidemics occurred in schools, hospitals and military camps, particularly when associated with institutional residences and crowding. Increased endemic infection was often noted at the same time.

- Symptoms: the symptom picture is characterized by its acute explosive onset, the severity of the CNS, autonomic and vascular symptoms, the associated malaise, and the often fleeting muscle and joint pains, but with a paucity of physical signs on physical examination and very low death rate. Little is said about duration and long-term findings. In my experience, depending upon the individual case, after weeks, months or sometimes years, this acute symptom picture gradually decreases in intensity. However, the average patient's intellectual, physical and emotional stamina, whose decrease tended to be noticed within 2-4weeks of illness onset, rarely recovers sufficiently for the patient to manage in the competitive world if the illness continues beyond 2 years. As in any true disease, there is variable severity and simply misdiagnosis.

- Pathology: the few deaths that went to autopsy demonstrated CNS injuries to the basal ganglia and other brain areas, and injuries to the anterior horn cells and dorsal root ganglia (A Chaudhuri, personal communication, February 2009). Unlike in paralytic poliomyelitis, the anterior horn cells tend to be injured rather than destroyed. Deaths occurred during and following the epidemics, including three children under 12years of age who died of Parkinson's disease within 2years of falling ill during the 1947 Iceland epidemic.

- Clinical tests: abnormal EEGs were found in the Royal Free epidemics[13,14]. Abnormal electromyography (EMG) was found in the LA, Copenhagen, Royal Free and Coventry epidemics. Neurologist Charles Poser found oligoclonal banding in some sporadic cases in referred patients at Harvard's Mount Sinai Hospital,[15] but he is perhaps the only doctor to have taken routine spinal fluids. Oligoclonal banding suggests CNS injury. These were all earlier epidemics. Outside of our own work, to my knowledge little if any systematic investigation has been done of significant groups of patients in the multitude of post-1984 cluster, epidemic and endemic cases.

- Cause: in the more than 60 epidemics and clusters described, only in four was an infectious source actually recovered and described. In four of the instances, this virus was an enterovirus or ECHO enterovirus, which were recovered in the later Akureyri episodes,[16] the Coventry epidemic[17] and the Ottawa 1984 clusters by DN Galbraith and C Nairn of Ruckhill Hospital, Glasgow (unpublished investigations). In Ontario, the provincial virologists also observed an association only with enteroviruses during the 1984-90 epidemic periods[18] and a negative association with Epstein-Barr virus (EBV). Concomitant gastrointestinal bacterial infection with Bethesda Ballerup paracolon group[19] was isolated in some of the patients in the Bethesda outbreak. It was not believed to be a cause. John Chia in California has recovered enterovirus from the gastric mucosa in multiple sporadic M.E. patients whom he has investigated.[20] The great difficulty in recovering polio enteroviruses in living patients should be remembered; almost all polioviruses in patients with polio were recovered from autopsy cases.

- Incubation period: in the 60 or so described epidemics, when an observed incubation period was noted, in most cases they were stated as 3-6 days.[21] Except for the 1934 LA epidemic, which was associated within days of immunization of the hospital staff, no other incidence of immunization association was specifically recorded. Most enteroviruses have an incubation period of 8-40 days, making EBV virus with a 40-day incubation period a highly unlike cause of epidemic M.E.

Definitions of CFS
These are definitions that started with the 1988 National Institutes of Health (NIH)/CDC Holmes definition.[6] The important fact with regard to this definition is that only 2 of the 16 authors had routinely investigated patients with M.E. and published on M.E. or CFS

before or after the publication. The 1988 definition was followed up by the 1991 Oxford Guidelines[22] and the 1992 NIH/CDC Fukuda definitions, which at best were copies of the 1988 definition. These three definitions are examples of theoretical definitions and are not based upon significant patient examination. If anything, they served to confuse the severity of M.E. that was associated with the term CFS. The later, so-called Canadian definition[8] is unique in the fact that the majority of the authors had long-term experience of examining patients with M.E. This definition discusses many of the findings in this chapter. The definition is lacking in that it was not based significantly upon actual organ and system pathophysiology and examination. It does not distinguish between M.E. and CFS. This definition is also much too long and complex; however, it remains the closest among the published definitions in describing the actual disease known as M.E.

The fault, dear reader, is not in our stars nor among the doctors or the patients, but in the system in which these two often apposing forces, the doctor and patient, theory and scientific investigation, exist. Until we actually do systematic long-term and in-depth scientific investigation of these patients, we will see no progress. Until then, I can only recommend that the doctor offers kindness to and toleration of these chronically disabled largely mistreated patients with M.E. and CFS. To the government of the UK, I can only recommend the nationwide funding of a serious scientific approach to the long-term pathophysiological investigation of these chronically disabled citizens.

And we 'remain' as on a darkling plain
Swept with confused alarms of struggle and flight,
Where ignorant armies clash by night.

KEY POINTS
• M.E. occurs as an epidemic and endemic disease. The injured patients as a group have rarely been investigated systematically during the past 50 years with the force of modern scientific and clinical investigational tools. Doctors investigating the patient with M.E. have

- rarely if ever received any significant funds for in-depth scientific investigation of organ and system pathologies.
- M.E. epidemic patients as a group have never been subjected to any long-term follow-up. Grufferman has suggested the possibility of increased cancer risk in the cohorts of epidemic M.E. -type patients.[23]
- M.E. and CFS are symptom-based definitions that have become simple garbage-bag terms for large numbers of patients with acute or gradual-onset physical and cognitive diseases affecting stamina, work and school ability. Thus, the terms M.E. and CFS have acted as an excuse to downplay the importance of these chronically ill patients and to not properly investigate these patients.
- The patient with M.E. has been largely disenfranchised from modern medical investigation and suitable medical assistance. Since most of these individuals tend to be relatively young, highly educated individuals in the medical and teaching professions, their loss of tax income to the government would more than offset their in-depth investigational costs.
- The majority of patients with M.E. or CFS that we have investigated in the UK, the USA or Canada represent multiple missed pathologies rather than any single disease entity. Whether these pathologies are caused by an initial CNS injury that deregulates the complex neuroimmune and neurochemical system and organ physiology of the patient or are simply co-morbidities is simply not known.
- There will never be a single treatment, whether pharmaceutical, physical, psychological or psychiatric, that will have any significant effect in the treatment of the majority of patients with M.E. and CFS, since they do not have a common trigger or common organ and system pathologies. It is necessary to first investigate the pathophysiological injuries and then treat them where possible to obtain a reasonable chance of a cure or treatment.
- The M.E. community continues to be a growing concern to many doctors and the state. This community is unlikely to disappear.
- Both patient and doctor mythologies continue to form a major part in the failure to assist the patient with M.E.

- One group of patients with M.E. and CFS for whom we know their pathology – the patients with dysautanomia – have seen no advance in funding or investigation of treatment in the past 30 years, to such an extent that most major cities and university medical schools in the UK, Europe and North America simply have no investigational ability in this area.
- The major problems in understanding the illness and disability of patients with M.E. and CFS lie in the following areas: (i) the Oslerian concept of a single pathology causing a single disease spectrum; (ii) the misguided notion that patients with M.E. and CFS are people who can 'think themselves sick'; (iii) the lack of funding and resources in the physical investigation into the pathologies and pathophysiologies of these patients; (iv) the failure to do any long-term follow-up study of this group of chronically ill patients; (v) the patient mythologies that magic treatments can cure the patient; and (vi) the erroneous and facile belief that patients with M.E. and CFS simply have variations of hysteria, somatization disorders or somatoform disorders in general.

REFERENCES

1. Parish JG (1992) A bibliography of M.E. /CFS epidemics. In BM Hyde (ed.) The Clinical and Scientific Basis of Myalgic Encephalomyelitis/Chronic Fatigue Syndrome. Ottawa: Nightingale Research Foundation.
2. Gilliam AG (1938) Epidemiological study of an epidemic, diagnosed as poliomyelitis, occurring among the personnel of the Los Angeles County General Hospital during the summer of 1934. Public Health Bulletin 240.
3. Sigurdsson B (1990) Ritverk: Collected Scientific Papers, 1936-1962. Reykjavik.
4. Hyde B (2003) The complexities of diagnosis. In LA Jason (ed.) Handbook of Chronic Fatigue Syndrome. Hoboken, NJ: John Wiley & Sons.
5. Hyde BM (1992) The definitions of M.E. /CFS: a review. In BM Hyde (ed.) The Clinical and Scientific Basis of Myalgic Encephalomyelitis/Chronic Fatigue Syndrome. Ottawa: Nightingale Research Foundation.
6. Holmes GP, Kaplan JE, Gantz NM, et al (1888) Chronic fatigue syndrome: a working case definition. Annals of Internal Medicine 108: 387-9.
7. Fukuda K, Stephen E, Straus SE, et al. (1994) The chronic fatigue syndrome: a comprehensive approach to its definition and study. Annals of Internal Medicine 121: 953-9.
8. Carruthers BM, Jain A, De Meirleir Kl, et al. (2003) Myalgic encephalomyelitis/chronic fatigue syndrome: clinical working case definition, diagnostic and treatment protocols. Journal of Chronic Fatigue Syndrome 11: 7-116.
9. Jason LA, Bell DS, De Meirleir K, et al. (2006) A pediatric case definition for myalgic encephalomyelitis and chronic fatigue syndrome. Journal of Chronic Fatigue Syndrome 13: 1-44.
10. Jason LA, Porter N, Shelleby E, et al. (2008) A case definition for children with myalgic encephalomyelitis/ chronic fatigue syndrome. Clinical Medicine: Pediatrics 1: 53-7.

11. Wickman I (1911) Die akute Poliomyelitis bzw: Heine-Medinsche Krankheit. Berlin: Springer.
12. Wickman I (1913) Acute Poliomyelitis: Heine-Medin's Disease (transl. J Maloney). New York: Journal of Nervous and Mental Disease Publishing Co.
13. Pampiglione G, Harris R, Kennedy J (1978) Neurological and electroencephalographic investigations in myalgic encephalomyelitis. Postgraduate Medical Journal 54: 752-4.
14. Warner C, Cookfiar D, Heffner R, et al. (1989) Neurological abnormalities in chronic fatigue syndrome. Neurology 39 (suppl. 1): 420.
15. Poser C The differential diagnosis between multiple sclerosis and chronic fatigue postviral syndrome. In BM Hyde (ed.) In BM Hyde (ed.) The Clinical and Scientific Basis of Myalgic Encephalomyelitis/Chronic Fatigue Syndrome. Ottawa: Nightingale Research Foundation.
16. Hyde B (1991) The Akureyi epidemic. In R Jenkins, JF Mowbray (eds) Post-Viral Fatigue Syndrome. Chichester: John Wiley & Sons.
17. Lyle WH (1959) An outbreak of disease believed to have been caused by ECHO 9 virus. Annals of Internal Medicine 51: 248-9.
18. McLaughlin MD (1991) Health and Welfare Canada: Proceedings of a Workshop. Ontario: Laboratory Services Branch, Ontario Ministry of Health.
19. Shelokov A, Habel K, Verder E, et al. (1957) Epidemic neuromyasthenia: an outbreak of poliomyelitis-like illness in student nurses. New England Journal of Medicine 257: 345-55.
20. Chia JK, Chia AW (2008) Chronic fatigue syndrome is associated with chronic enterovirus infection of the stomach. Journal of Clinical Pathology 61: 43-8.
21. Hyde B, Bastien S, Jain A (1992) Patterns of neuropsychological abnormalities and cognitive impairment in adults and children. In BM Hyde (ed.) The Clinical and Scientific Basis of Myalgic Encephalomyelitis/Chronic Fatigue Syndrome. Ottawa: Nightingale Research Foundation.

22. Sharpe MC, Archard LC, Banatvale JE, et al. (1991) A report: chronic fatigue syndrome - guidelines for research. Journal of the Royal Society of Medicine 84: 118-21.

23. Grufferman S (1992) Epidemiologic and immunologic findings in clusters of chronic fatigue syndrome. In BM Hyde (ed.) The Clinical and Scientific Basis of Myalgic Encephalomyelitis/Chronic Fatigue Syndrome. Ottawa: Nightingale Research Foundation.

FIGURES

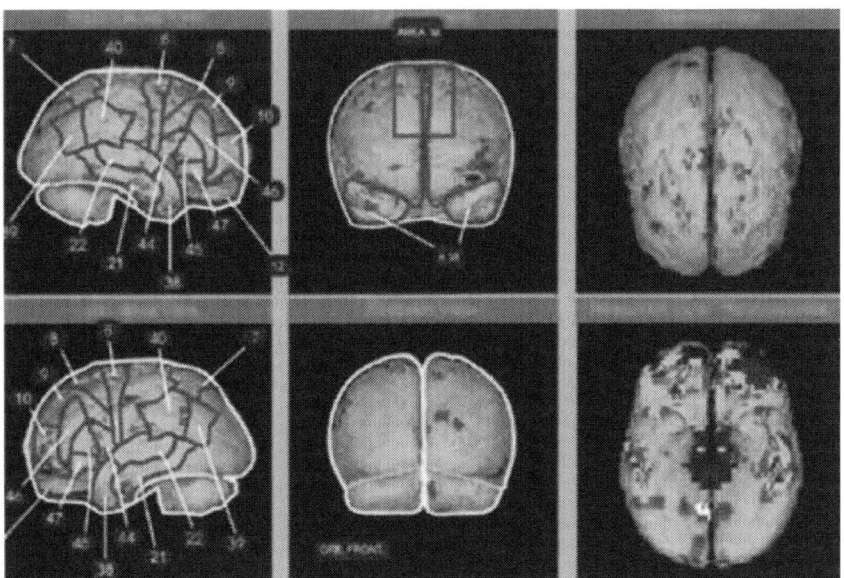

Figure 51.1 Brain single-photon-emission computed tomography (SPECT) of typical patient with myalgic encephalomyelitis (M.E.). This is a typical severe M.E. brain, as visualized by SPECT at illness onset, with Brodman areas indicated. Blue areas represent the significant typical hypoperfused areas of this dysfunctional brain. This schoolteacher fell ill following a severe influenza-like infectious episode 1 week after return to school. The patient went on to develop thyroid malignancy and Clostridium difficile infection several years later. Typical M.E. features: teacher, female, age 40 years at illness onset, post-infectious, no significant recovery during 15-year follow-up.

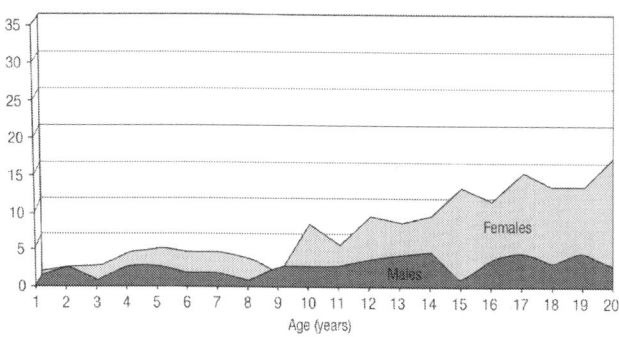

Figure 51.2 Sex and age of patients with myalgic encephalomyelitis (M.E.)/chronic fatigue syndrome (CFS) seen between 1984 and 1992 at the Nightingale Research Clinic: children and youths (age 0-21 years)

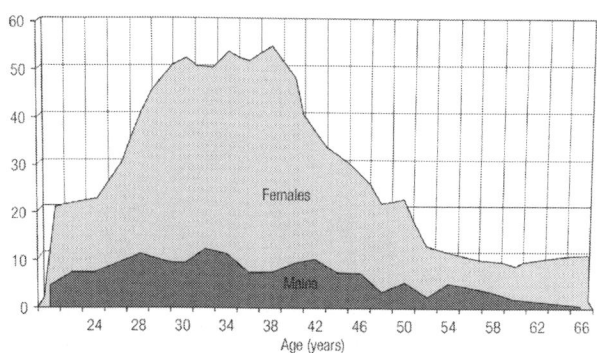

Figure 51.3 Sex and age of patients with myalgic encephalomyelitis (M.E.)/chronic fatigue syndrome (CFS) seen between 1984 and 1992 at the Nightingale Research Clinic: adults (age 22-70 years)

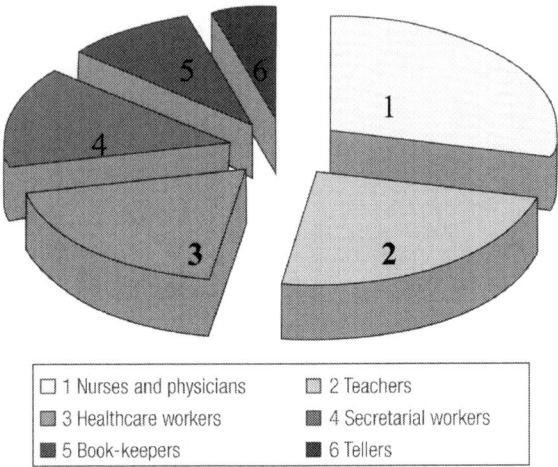

Figure 51.4 Occupation of patients with myalgic encephalomyelitis (M.E.) at illness onset

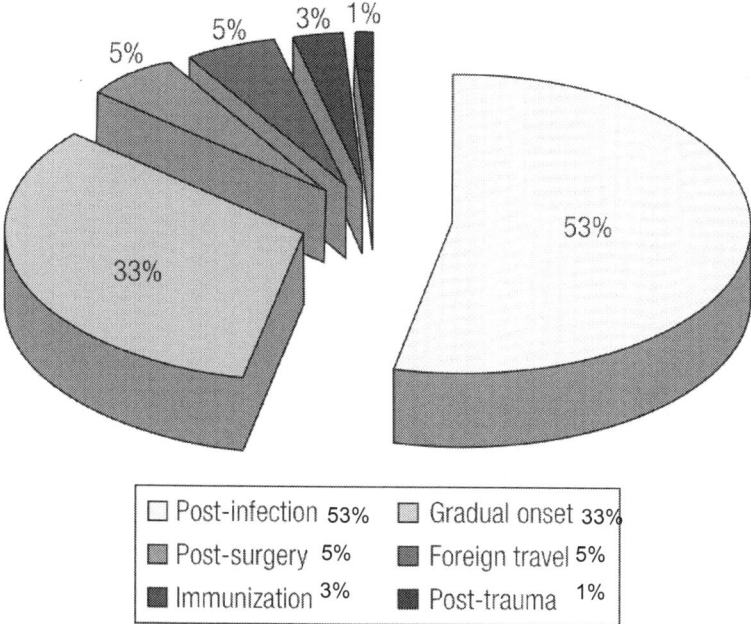

Figure 51.5 Triggers ascribed by the patient as a possible cause of their present illness in 2000 cases of myalgic encephalomyelitis (M.E.)/chronic fatigue syndrome (CFS), 1984–1992. The gradual-onset patients tended to have no apparent associated triggers

INDEX

atherosclerosis, 44
autonomic, 18, 19, 22, 29, 48, 54, 80
Babinski, 36
basilar artery, 45, 60
body mapping, 11, 24, 25
Bowel Dysfunction, 82
brain, 10, 11, 18, 27, 32, 42, 45, 48, 49, 50, 52, 53, 54, 55, 60, 61, 78, 80, 82, 88, 91, 92, 93, 95, 96
brain atrophy, 52
calcification, 53
carcinoma, 40
cardiac, 18, 19, 22, 29, 40, 47, 53, 55, 56, 61, 82, 83, 85, 92
Cardiac Irregularity, 81
CFS, 5, 9, 10, 11, 12, 16, 19, 20, 21, 22, 23, 24, 25, 26, 27, 28, 29, 30, 31, 32, 34, 35, 36, 37, 38, 39, 40, 41, 42, 43, 44, 45, 46, 47, 48, 49, 52, 53, 54, 56, 57, 58, 60, 61, 62, 64, 69, 71, 72, 73, 74, 77, 79, 85, 87, 88, 89, 90, 92, 94, 95, 97, 100
CHRONIC FATIGUE SYNDROME, 1, 9, 10, 13, 16, 64, 70, 72, 87, 92, 99, 100
Circulating Blood Volume, 81

Central Nervous System, 17, 18, 19, 20, 21, 22, 29, 33, 39, 44, 49, 50, 52, 53, 55, 60, 61, 75, 77, 78, 79, 80, 88, 90, 95
cognitive, 27, 78, 94
Crohn's disease, 60
diabetes, 37, 58, 80, 86
Doppler, 43, 44, 45, 51, 53, 55, 56, 60, 96
Dowsett, 87, 90
echocardiograms, 43
EEG, 54, 55, 96
Ehlers-Danlos Syndromes, 82
Endocrine, 85
enterovirus, 42, 60, 61, 86, 90
epidemic, 5, 17, 19, 20, 22, 42, 63, 65, 72, 75, 76, 77, 88, 89, 99, 100
epidemics
  epidemic, 9, 17, 54, 64, 69, 72, 74
Factor V Leiden, 84
fibromyalgia, 9, 28, 29, 36, 47, 50, 65, 77, 95
Fukuda, 20, 21, 22, 63, 72, 99
glucose, 37, 38
Goldstein, Dr. Jay, 16, 28, 49, 62, 63, 64, 87, 88, 93, 99, 100
gut, 82

Hashimoto's thyroiditis, 39, 59, 85
Hughes Syndrome, 84, 99
Hyde, Byron Marshall, 1, 2, 5, 9, 10, 11, 12, 13, 16, 17, 52, 53, 63, 64, 72, 81, 87, 92, 99, 100
immune, 19, 30, 32, 39, 40, 59, 61, 76, 77, 91
immunization, 9, 18, 19, 38, 59, 76, 88, 90
inflammatory, 37, 95
insurance, 28, 30, 31, 41, 51, 52, 57, 59, 73, 92, 94
investigation, 10, 11, 16, 23, 24, 25, 26, 27, 28, 30, 40, 41, 53, 56, 59, 60, 65, 74, 89, 91, 92, 94
Jason, Dr. Leonard A., 87
Kerr, Dr. Jonathan 90
lawyer, 41
leukemia, 60
lumbar puncture, 41
lymph nodes, 19, 34, 35
M.E., 2, 3, 5, 16, 17, 18, 19, 20, 21, 22, 23, 24, 25, 26, 27, 28, 29, 30, 31, 32, 33, 34, 35, 36, 37, 38, 39, 41, 42, 43, 44, 45, 46, 47, 48, 49, 50, 52, 53, 54, 55, 58, 59, 60, 61, 62, 64
malignancy, 11, 29, 35, 36, 37, 38, 40, 52, 55, 61, 85, 91
Mena, Dr. Ishmael, 49, 62, 64, 87, 88, 93

Mowbray, Dr. James, 87, 90
MRI, 36, 49, 51, 52, 53, 95
MYALGIC ENCEPHALOMYELITIS, 1, 2, 9, 10, 16, 67, 69, 70, 71, 72, 73, 76, 78, 93, 99, 100
neuro-endocrine, 75
neurologist, 54, 62
neurophysiological, 78
Neuropsychologist, 78, 92
neuroradiologist, 36, 51
Newcastle Research Group, 10, 87
Nightingale Research Foundation, 9, 16, 63, 64, 65, 72, 97, 99, 100
Paediatric, 9
Parathyroid, 38
Parish, 35, 87
pathology, 22, 26, 27, 29, 32, 34, 39, 40, 43, 44, 45, 46, 47, 51, 53, 54, 56, 62, 80, 88, 91, 94, 95, 97
pathophysiological, 23, 24, 74
pelvic tumors, 46
pericardial, 43, 44
pesticides, 32, 91
pharmaceutical, 21, 93, 95
physicians, 2, 9, 10, 11, 12, 16, 21, 22, 23, 24, 26, 28, 30, 31, 33, 37, 39, 41, 42, 44, 46, 47, 49, 51, 52, 54, 56, 57, 58, 60, 69, 70, 71, 72,

73, 74, 81, 82, 84, 86, 87, 88, 90, 93, 94, 95, 96
pituitary, 40, 52, 80
plantar reflex, 36
polio, 31, 43, 86, 89, 90
Poser, Dr. Charles Poser, 62, 87
posterior root ganglia, 28, 75
POTS, 80, 83, 84, 86
Protein C, 84
Protein S, 84
psychiatric, 21, 23, 24, 26, 32, 56, 57, 58, 62, 72, 73, 92
QEEG, 55, 78, 96
Ramsay, 17, 64, 72, 87, 90, 100
Raynaud's Phenomenon, 81
rheumatoid, 37, 38, 39, 81, 95
Richardson, Dr. John, 10, 17, 19, 29, 43, 44, 62, 64, 86, 87, 90
Romberg, 36
Sigurjonsson, 65, 100
skeletal, 82
sleep dysfunction, 53, 54, 94

SPECT, 10, 28, 36, 39, 42, 45, 48, 49, 50, 59, 60, 61, 64, 78, 79, 80, 88, 91, 93, 96
spinal fluid, 28, 41
Streeten, David, 80, 101
stress
   stressor, 20, 29, 46, 47, 50, 55, 83
subcortical, 18, 28, 45, 48, 50, 79, 82, 89
subcutaneous, 35
temperature, 22, 34, 38, 81
tests, 22, 23, 26, 27, 30, 33, 36, 37, 38, 39, 40, 41, 42, 43, 46, 47, 48, 49, 51, 53, 56, 58, 59, 73, 78, 79, 84, 90
Thyroid, 11, 38, 43, 85, 91
ultrasound, 38, 39, 40, 43, 46, 59, 86, 96
upper motor neuron disorders, 33
vasculitis, 50, 59, 61, 83, 88
viral, 42, 59, 61, 63, 82, 83, 89
Wessely, Prof. Simon, 5, 20, 64
Zenon, 93

Printed in Great Britain
by Amazon.co.uk, Ltd.,
Marston Gate.